A Roadmap for Curing Cancer, Alzheimer's, and Cardiovascular Disease

A Roadmap for Curing Cancer, Alzheimer's, and Cardiovascular Disease

Paul J. Marangos

ELSEVIER

Elsevier
Radarweg 29, PO Box 211, 1000 AE Amsterdam, Netherlands
The Boulevard, Langford Lane, Kidlington, Oxford OX5 1GB, United Kingdom
50 Hampshire Street, 5th Floor, Cambridge, MA 02139, United States

Notices
Knowledge and best practice in this field are constantly changing. As new research and experience
broaden our understanding, changes in research methods, professional practices, or medical
treatment may become necessary.

Practitioners and researchers may always rely on their own experience and knowledge in
evaluating and using any information, methods, compounds, or experiments described herein.
In using such information or methods they should be mindful of their own safety and the safety
of others, including parties for whom they have a professional responsibility.

To the fullest extent of the law, neither the Publisher nor the authors, contributors, or editors,
assume any liability for any injury and/or damage to persons or property as a matter of products
liability, negligence or otherwise, or from any use or operation of any methods, products,
instructions, or ideas contained in the material herein.

Library of Congress Cataloging-in-Publication Data
A catalog record for this book is available from the Library of Congress

British Library Cataloguing-in-Publication Data
A catalogue record for this book is available from the British Library

ISBN: 978-0-12-812796-4

For information on all Elsevier publications visit our website at
https://www.elsevier.com/books-and-journals

  Working together
to grow libraries in
developing countries

www.elsevier.com • www.bookaid.org

Publisher: Mara Conner
Acquisition Editor: Natalie Farra
Editorial Project Manager: Kathy Padilla
Production Project Manager: Julia Haynes
Designer: Miles Hitchen

Typeset by TNQ Books and Journals

*To the children and grandchildren of the world
so they may be free from terminal disease.*

Contents

Part B
Towards Solutions

x   Contents

Preface

I have spent the past 45 years in the medical R&D arena. The first 14 years at the Roche Institute of Molecular Biology and the National Institutes of Health (NIH) doing basic research and 31 years in the drug industry cofounding and leading five different companies. I have therefore traveled through every corridor of the system charged with cures in search of something to pin my ambition to ridding us of the terminal disease nightmare and been universally frustrated. There is no logic to any of it, and therefore the only solution I see is to write this book. I and my colleagues, many much smarter than me, have failed in the arena of cures. We were pushing against a wall of fog and until that changes the effort borders on futility.

The failure to cure terminal diseases is the greatest global threat to humanity. It dwarfs any other challenge we face as a society, and yet it was not a major issue in the recent election. The delivery of healthcare was center stage as is appropriate, but from the standpoint of saving human lives it does not approach the importance of curing the terminal diseases. Cancer, degenerative diseases such as Alzheimer's and Parkinson's, and cardiovascular disease prematurely end almost 2 million lives every year. What other threat to humanity approaches this magnitude? If we took curing terminal diseases as seriously as providing healthcare, we would very likely have been successful long before now. No one has dared to critique the entire system to determine why it has failed so dramatically but rather they just keep throwing more dollars at it. Broken is too kind a word to describe it since it is not a coherent entity; criminal negligence comes a lot closer to the reality.

We are playing parlor games with cures. The fault is not in us but rather in a system that was never designed for the task. The talent is there as well as the capital and infrastructure. The libraries are flooded with research papers so productivity is not an issue but yet the goal is apparently absent. The millions of research papers have served science and its practitioner's well but have clearly failed when it comes to results. We are buried in science but cures seem to be an afterthought. The roadblocks we have created at almost every step of the process have made drug development a billion-dollar decade-long Sisyphean task. History clearly shows what we are doing has not worked and what follows will point out why and specify a course of action that will put the cure quest on a laser-focused course to finally getting the job done.

We have witnessed quantum technological advances during the past several decades. Virtually every other aspect of our lives has been transformed over this period but when it comes to the most important of all challenges—terminal diseases cures—we have failed. Also, those past successes were made without the help of the super data management and communication capabilities we have had for the past several decades, so it is rather bizarre that the cures mission has failed so dramatically and did not justify the coordinated focused effort given to the conquest of space travel, cyberspace, and computing. But rather the "effort" was allowed to evolve into an academic exercise and be strangled by regulation and legal assaults. Having a supercomputer in your pocket, a nuclear power plant down the road, and folks orbiting in a space station is fine, but if terminal diseases were conquered that would be orders of magnitude more significant. This is not a simple task but neither were those other successes.

Curing terminal disease is left to the whim of researchers at universities, institutes, and charities all proceeding to leisurely pick away at basic science in a totally unaccountable disorganized fashion in hopes of stumbling on something that might be relevant to these diseases. While the NIH is out there as the federal government's gesture to health research, it has yet to make major inroads into the terminal disease crisis after almost a century of supposedly trying. It consists of 27 institutes, which seems ridiculous given that just three disease classes take the great majority of all lives, and nowhere in its mission statement is curing terminal disease or any disease mentioned. We have engaged in science for science sake for the past 50 years and until that ends very little will change.

Another fundamental reason for the failure in the medical arena is that the entire development component is regulated to the point of strangulation by the Food and Drug Administration (FDA) and encumbered by a patent and legal establishment that makes drug development extremely expensive and difficult. Terminal patients face certain death, yet the regulators insist on usurping their right to life by practicing an arrogant academic purism that has held up drug approvals, made access to investigational drugs very difficult, and transformed the price of drug approval into a several billion dollar venture. Until the regulators and legal system back off substantially regarding terminal disease therapeutics, there will continue to be a very slow progress.

The problems in specific areas of the medical R&D establishment have only recently begun to be appreciated but no one has put the entire establishment under a microscope in one place and proposed solutions; that is the reason for this book. Once the reasons for failure are exposed and reforms proposed that will speed this seemingly endless waiting game for cures, the hope is that the public will rise up and demand action. Unless this is made an issue by the electorate nothing of any consequence will happen. The stakeholders will smother it, especially those who actually benefit from the status quo, which is just about all of them. I am proposing substantial changes in virtually every aspect of the system and there will be kickback. Congress will have to override that but if the public makes it clear that how they vote will be determined by what the congress mandates then this will happen.

The AIDS situation is a good example of what can happen if the public rises up and makes it clear that they want results. The AIDS crisis was championed by a small group of sophisticated individuals that took to the streets, marched into the halls of government, and demanded action. Washington got the message loud and clear. The NIH went into high gear and identified the virus rapidly as did the Pasteur institute. The FDA stepped out of the way and accelerated the approval of the drug cocktail that transformed AIDS from a terminal disease to a treatable one thereby saving thousands of lives. Unfortunately that mentality did not stick because the FDA was not organically changed regarding its mission, management, and attitude about a patient's right to life as I am proposing. The NIH was also not reformed as is urgently required.

Providing the ammunition to start a movement to get serious about the cure crisis is a major goal of this book. We cannot be satisfied with millions dying every year while researchers proceed to amble through their individual research ambitions; drug companies have no incentive to develop cures; the NIH and academia proceed in an unfocused and unaccountable manner; the FDA strangles the drug industry; a patient's right to life is destroyed and the patient–physician relationship disappears. If we fail to make a serious effort to reform the cure effort and continue to pour money into a failed system, the road to cures will be a very long one. These issues have percolated in the political arena but it has always been about more money to build more laboratories and never about how we craft a global effort to achieve success. It is not about money but rather about strategy and repurposing what already exists.

This book seeks to raise awareness about the folly that is the medical R&D enterprise. My primary target audience is the public and providing it with the knowledge necessary to fight this battle. It will be a battle because the status quo and vested interests need to be brushed aside since they have clearly failed. The book is critical of virtually every corner of the enterprise; academia, the drug and biotech industry, thought leaders, venture capitalists, charities, the FDA, the patent codes, the legal system, the healthcare delivery system. To accomplish the task the discussion will be in two parts. The first dissects the drug development establishment and points out reasons for the paralysis of progress in plain nontechnical terms. The second part suggests solutions that if appropriately thought out and put in place will transform the current quest for cures into an organized, accountable effort with the focus, urgency, and strategy required for success.

For decades we have all watched loved ones die needlessly and live in fear of terminal disease. Cancer and Alzheimer's are two of the worst experiences a family can face. One kills the body, the other the mind, and yet we seem pacified by sporadic reports of science breakthroughs that never lead to anything tangible aside from Nobel Prizes and prestigious institutions immune from accountability. It does not have to be this way if we finally get serious about ending the nightmare by applying our magnificent abilities in the same manner that produced the wonders that have transformed every other aspect of our lives.

Acknowledgments

I would like to thank the Wall Street Journal for their reporting on the medical R&D system over the past several decades. They repeatedly raised issues that I had experienced during my career. They did a front page story over a decade ago[1] on the search for a cure to Alzheimer's disease. The piece, written by Sharon Begley, focused on the fact that for the past 30 years only one approach to the problem was being taken, and that it was not working. The author questioned the logic of this, and it immediately struck a chord with me because I had wondered the same thing for decades. On further thought it became apparent to me that this huge mistake in attacking Alzheimer's disease was due to the fact that there was no strategy, accountability, or urgency involved in the effort, and that this was also true for the other terminal diseases. The book grew out of that excellent reporting.

In the ensuing years the Journal has consistently raised awareness on the problems that have impeded curing terminal disease and they are heavily, but not inclusively, cited in the book. I thank them for this public service and commend them for excellent reporting. If this book accomplishes its purpose of enabling cures for terminal diseases and ending this terrible human tragedy, much of the credit should accrue to the Wall Street Journal.

---

1. S. Begley, Is Alzheimer's field blocking research into other causes? The Wall Street Journal (April 9, 2004). Retrieved from: www.wsj.com/articles/SB108145279348578177.

Part A

The Reasons for Failure

Chapter 1

The Medical R&D Fiasco

Chapter Outline

THE MOST SERIOUS HUMANITARIAN CRISIS

Everyone has experienced the tragedy of the three major terminal diseases: cancer, Alzheimer's disease, and cardiovascular disease. We watch helplessly as either the life or mind of a loved one fades away. In my case it was my mother, three grandparents, friends, and relatives. How long must we tolerate this? The older one becomes, the more frustrating this is as the decades pass and nothing changes. All we hear about for the past 40 years are the science breakthroughs, none of which has made a major difference. In spite of the 1971 "War on Cancer" and multiple spikes in the National Institutes of Health (NIH) annual budget over that same period, terminal diseases still claim just shy of two million lives every year! This global threat to humanity dwarfs any other and yet we relegate it to charities, the whims of academia, and a rudderless effort lacking any reasonable degree of strategy, accountability, or urgency. Add to that the inane regulations, which make drug development almost impossible and enormous class action lawsuits that cripple, and in some cases, destroy companies and it becomes clear why so little progress has been made.

We have conquered outer space, cyberspace, computing, and split the atom yet we cannot seem to get to first base when it comes to curing a single terminal disease. We have been in a cure crisis for the past 60 years and are seemingly helpless to end it. It is now almost accepted as part of life. This global threat to humanity dwarfs all others including war, climate change, famine, or the delivery of healthcare. Some have written about this and expressed frustration [1–3], but the remedy has been elusive. The entire system [National Institutes of Health (NIH), Food and Drug Administration (FDA), academia, the drug

A Roadmap for Curing Cancer, Alzheimer's, and Cardiovascular Disease.
http://dx.doi.org/10.1016/B978-0-12-812796-4.00001-5

industry, legal system, charities] is broken and must be changed in its entirety at the most fundamental level. The good news is that the reforms required to convert this effort from a parlor game to a serious assault require minimal new infrastructure or enhanced investment.

The discussion that follows will be in two parts. Part A will systematically reveal the reasons for this colossal failure as well as inform the reader on the nuts and bolts of the entire medical research and development (R&D) enterprise. Understanding the system and the basic issues the three terminal diseases present is critical to solving them. Part B will propose pragmatic and doable solutions for each sector of the R&D system so that the public can advocate and demand that the stakeholders and governments take the necessary action to finally end the carnage.

THE MAGNITUDE OF THE PROBLEM

The entire medical R&D arena has failed. While it is true that advances have been made in combating the terminal diseases, they are largely attributed to changes in lifestyles such as smoking cessation, dietary improvements, and early detection imaging diagnostic tools and not major advances of cures. The American Cancer Society predicts that there will be 595,690 cancer deaths in the United States in 2016. The Alzheimer's Association reports 84,000 deaths for Alzheimer's patients in 2013 and projects that the number will increase dramatically when 2016 figures are available. Keep in mind that memory dies far more rapidly than the body in this disease so the actual toll of Alzheimer's is probably more like 500,000 lives essentially ended each year. Some progress has been made with treating cardiovascular disease (heart attacks and stroke and related disorders), but according to the American Heart Association there are still 800,000 deaths per year in the United States and over 17 million worldwide directly related to cardiovascular disease.

When totaled, in the United States more lives are lost in just 1 year (about 1.9 million) from the three major terminal diseases than in all the wars the nation has had in its entire history. This is the largest threat to humanity that ours or any country faces, and we do not treat it as such. Unfortunately, we have never made cures a national priority but have rather mindlessly pumped hundreds of billions into organizations not designed to find cures such as universities and research institutes. We have put unwarranted faith into charities that seemingly do a good job of creating solidarity among those affected with these diseases but have produced virtually nothing on the cure front. These efforts have been devoid of strategy, coordination, and accountability, the key ingredients that led our nation to some of its greatest accomplishments. We have also tolerated a regulatory and legal environment that impedes the companies charged with solving the problem. Amazingly, the people have tolerated this and for decades have watched their loved ones unnecessarily suffer.

Once the public sees chaos we call medical R&D in its entirety, and that it can be fixed to accelerate cures, the path to reform will be clear. It has to be a grassroots uprising that demands fundamental change and not just injecting more money, which has repeatedly failed. The reason is that there will certainly be resistance to the reforms called for from the stakeholders with vested interests who all are quite happy with the gravy train the status quo has produced. It is appropriate to begin the critique of this chaotic system summarizing my experiences and observations over the past 45 years that has led me to this conclusion. Experience is the best teacher because it involves no intermediary with their translational bias.

MY JOURNEY THROUGH THE QUAGMIRE

In 1973, I had just received my PhD in biochemistry, a discipline I truly loved both for its elegance and relevance to the human condition. The biochemical choreography of life was and remains fascinating to me. DNA is the blueprint of life protected in the chromosome residing in the nucleus of the cell. DNA guides the synthesis of messenger RNA, which travels to the cytoplasm of the cell where with the help of an assembly line called a ribosome and transfer RNA meticulously links amino acids together to form the proteins that conduct the business of life. I loved everything about this symphony of life as well as the university training that introduced me to it. Now I had to decide what to do with it. I chose to pursue a postdoctoral fellowship, which for a PhD is analogous to an internship in medicine, in neuroscience, and more specifically neurochemistry. I felt at that time that brain diseases were the last frontier in medical research.

Researchers brimmed with optimism in the mid-1960s that breakthroughs were on the horizon for many terminal diseases, especially cancer. We knew what the cancer cell was and how it differed from normal cells and that convinced me back in 1973, just 2 years after the war on cancer, that cancer would soon be cured. Regarding cardiovascular disease, we knew that terminal heart disease and stroke was simply a plumbing problem with clogged arteries starving the body of oxygen. I was therefore certain that cancer and cardiovascular disease would be cured in short order. So, I decided to set my sights on neurodegenerative disorders since we knew a lot less about brain diseases. I believed that breakthrough treatments and cures for the neurodegenerative diseases would be forthcoming during my career. Surely my generation, and certainly that of my children, would be free from these frightening conditions, and I could play a part in that. Little did I know how wrong I was?

By 1976 I had completed my postdoctoral fellowship at the Roche Institute of Molecular Biology in Nutley, New Jersey and had published a number of scientific papers in prestigious biochemistry and neurochemistry journals. Our knowledge of how the brain worked was growing rapidly, and the thought leaders preached that we were hot on the trail of cures for the neurological diseases as well. The seeds of doubt were, however, beginning to germinate, even at this seminal stage of my career. I was becoming disillusioned by the academic approach to disease-related research, and how disorganized and unfocused it was. There was no accountability or strategy to cure anything but rather an infatuation with publishing new science for its own sake and winning a Nobel Prize. With a growing disillusionment, I saw that most medical research was not about curing disease, but about academic freedom, awards, publishing, and tenure. It seemed to me that academics cared more about posing questions that would attract grant funding and less about solving real-world problems. Adoration by your peers and winning prestigious research awards and tenure were the pots of gold at the end of this rainbow. I remember a professor bragging about the fact that he had tenure now and could do whatever he wanted. The majority of academics I knew cared more about how many publications they had than how many problems they solved; it was all about building a long bibliography. I was drawn into this publish- or-perish frenzy when it became obvious to me that this was the most important criteria for advancement. My early misgivings about misplaced priorities were therefore put aside and I jumped in to play the game by the rules that prevailed.

My first real job was at the NIH, a bastion of free thought and generous research funding. I did not have to submit grant applications but rather could follow an idea the day I got it. That to me was nirvana, and I focused my research on finding ways to prevent or minimize brain damage following a stroke or head trauma thinking that this was an easier way to study how to protect the brain when it was injured by an acute event such as a stroke. I was fascinated by the prospect of reducing the tragedy of brain injury and the possibility that such treatments might also stop the progression of the chronic neurodegenerative diseases such as Alzheimer's disease. The NIH was the ideal place to engage in such a quest—at least so I thought at the time.

In 12 years at the NIH, I published hundreds of original scientific papers, edited several books, founded "The Journal of Molecular Neuroscience," and earned research awards. This all served my academic ego and cures were irrelevant, the bibliography, however, had grown to 30 pages without including abstracts or invited lectures, and it got me tenure in short order. The rush to publish and gain the respect and acceptance of my peers was what I was chasing because that was what rewarded by the establishment. It felt good to be successful at that. My initial goal was no longer relevant to advancing my career but rather would delay it. When I realized this, I became uncomfortable with the academic culture at the NIH and even more so at the universities at which many of my collaborators worked.

At the NIH it was impossible to translate my work into new treatments since the specific resources and expertise for doing so were not available. These were considered mundane tasks. While there were chemists on site who could make the drugs I thought would be interesting to test, they were off on their own chemistry research projects and had little interest in making what I wanted. They were more interested in discovering a new chemical reaction and not making a drug by standard methods, which was not new science but purely manufacturing. The many other support functions required for drug development were also not available at the NIH since they did not involve research, but rather the mundane activities involved in product development. New science was what the journal editors wanted, and anything with a commercial component was viewed as not being on the frontiers of science. In a nutshell, applied research was difficult to do and viewed as mundane science. However, applied research is what drug development is about. When I tried to collaborate with industry in this regard, it was very difficult and required numerous approvals and restrictions both from the NIH and the companies. The NIH did not want to be put in a conflict situation and the company needed patent protection to move forward, something that was not a priority for me since it impeded publication.

By the mid-1980s a cancer cure was still nowhere in sight. Patients were still being treated with the cytotoxic chemotherapy drugs used in the 1960s. The situation was much worse for Alzheimer's disease and all the other neurodegenerative diseases. While researchers were discovering important mechanisms in the laboratory, their animal models did not translate well to human disease mitigation. The data in animals was difficult to reproduce, and the lack of quality control was obvious. The rush to new technologies was trumping the application of existing science (applied research) to cures. There was so much data out there that was neglected such as last-year's car models. No one was interested or motivated to engage in the repugnant practice of applying old science but rather to forge ahead and generate more science to fill journals and build resumes, and I was among them with a resume that boasted several hundred peer-reviewed papers but no closer to anything resembling a new treatment or cure.

Everyone at the NIH, me included, was programmed to follow what was in vogue at the time rather than to be mavericks and look off the beaten path for new insights on disease breakthroughs. This was an easier path to the tenure covenant. Creativity and innovation were channeled to politically correct science dictated by the "thought leaders" who were the Nobel Laureates or the "prestigious" institute directors and department heads. Anything commercial or applied was an insult to the purity of science and a lower form of endeavor. Industry was viewed as the enemy of academic freedom. There were few mavericks, something I would later come to recognize as being critical to solving elusive problems that were resisting mainstream approaches. I had received my tenure at NIH by 1982 and was well on my way to an international reputation, so I was set for life but not happy with myself.

By this time my disillusionment with the entire academic mentality was palpable. I felt I was not doing much more than filling journals with data, virtually all of which lacked any formal quality control and sometimes turned out to be difficult to reproduce in the rush to publish. We generated our data in the laboratory and analyzed them ourselves in an unblinded fashion. This is an unacceptable practice and one not acceptable to the regulatory bodies that evaluate industry data. Strict quality control procedures are required by the FDA when evaluating drug company data. Also, journal editorial boards are generally not schooled in quality control procedures, more on that later. I needed more tangible real-world results from my efforts, a strategy, a timeline, and an end in sight that had direct impact on curing disease. Science for its own sake, while attractive, seemed ridiculous to me given the millions dying each year. Perhaps the business world would be a more appropriate place to get something tangible done.

In 1988 I fled the NIH, tenure and all, to join a small start-up biotech firm in Southern California that had just received its first round of funding. I went from the most secure employer on the planet, the US government, to a company with only 2 years of capital on hand and just an idea. It was so exciting. The company's mission was to develop an innovative new drug to protect the heart and brain during open-heart surgery and stroke. Finally I would be working on something tangible, a drug that could treat a problem, which was then, and still is today, untreatable. This is what all medical researchers should be doing given the urgency of the terminal disease crisis, so why did it take me so long to wake up to that? It felt like being reborn, an epiphany, or both.

I found the start-up business world invigorating with a crystal clear mission, urgency, and accountability. My company had a plan, a clear goal, and a timeline to develop new treatments that would save lives rather than filling journals with data of questionable quality and impress peers with its brilliance. The stakes were much higher and failure was not an option. In the academic world a negative result simply meant asking a different question. In the start-up drug development business, if the drug fails, the company usually goes under and you lose your job. There was also a greater degree of intellectual honesty about goals. The development of new and better drugs at a good return on the invested capital was the pure purpose, not veiled ego massaging or vain pursuits of academic accolades. Management prevailed, something that was foreign to me. Project teams analyzed every aspect of a development program and were assigned appropriate resources to get the work done. Data was checked and audited to insure quality control, something that is never done in academia. A third party, the FDA, audited the company's raw data to make sure the company was not biased in its interpretation, something that would insult academic scientists. There was strict accountability; when people did not perform, they were replaced, from the CEO on down. Tenure did not exist and if it did it would have insured failure. This was the real world where bottom-line results ruled the day.

I found the two and a half years spent at this start-up firm to be transforming, so much that I left and ventured out on my own and formed my own company. This was very difficult to do with several preteen children and a big mortgage, but my wife said do it and I was on my way to the new world of putting together a start-up biopharmaceutical company, raising money, writing a business plan, a 5-year profit and loss statement, and literally learning by doing. I brought in some business expertise with a cofounder and was off and running. I put some of my own drug development research ideas into the new company and realizing that I needed some later stage projects to get the company funded. I licensed the patents to two compounds that had already been tested in human subjects. One showed promising results as a protective agent for brain injury and the other for cardioprotection. This was even more exciting and stressful than leaving the NIH since I was without a job for 16 months getting the licenses and raising money to fund this new company. An Angel investor put it the first money followed by others. An eventual initial public offering and additional capital raises enabled bringing both licensed drugs into late-stage clinical trials, but regulatory hurdles and a downturn in the investment arena prevented these promising products from ever reaching patients.

The process of getting a drug to market was just too cumbersome and cost intensive. We did, however, manage to salvage the company by changing it to a specialty pharmaceutical company, which acquires marketed drugs to generate revenue streams. The company did achieve a degree of success as a specialty pharmaceutical business and was eventually, after a merger with another start-up company it was purchased by a larger pharmaceutical company, but it produced no cures for terminal diseases. The success was purely financial and not what I had started out to do so it was fundamentally a disappointment for me. The regulatory hurdles proved insurmountable and the experience demonstrated just how deep the problems were with getting any new drug approved and how inefficient start-up companies were.

During this period I became very familiar with virtually every aspect of drug development such as: the FDA, clinical trials, intellectual property strategies, product sales and marketing, and the capital formation to support it all. I now had a global view of both the research and business issues of new drug development, and what I saw was not good. Getting a drug approved was just as frustrating as mobilizing academics to apply science rather than do it pursue their academic freedom. It now became clear to me that the entire R&D system was fundamentally flawed when it came to developing cures. It would be necessary to scale back my aspirations for future companies to be successful from a business perspective and to forget about cures. One has to eat.

I proceeded to cofound four additional companies, all of which were more restricted in scope as I came to realize how hopeless it was to get new therapies to patients in my lifetime. These later companies dealt with *improving* existing drugs through employing novel reformulations, not *developing* new ones. While the companies achieved varying degrees of business success, they were not designed to achieve the goals I initially dreamed of 30 years earlier. The industry had as many impediments to developing cures as the academic community had in discovering them.

The progress that my colleagues and I had made during 3 decades in the drug development business was undeniably dismal, and the uniformity of our experience made it unlikely that it was our fault. We could not all be wrong since we were all hitting the same brick walls. The real tragedy was that these systemic obstacles were killing two million people per year in the United States and bankrupting the healthcare system. I therefore began to think there was only one course to pursue in the twilight of my career. The arena for finding cures had to be established. There is no road map, no organized effort, no accountability, and above all no urgency. I needed to write a book.

It is important to understand that during my journey, my initial feelings about each aspect of the system were positive. I loved the university campus and academic life. The start-up company was wonderful and the entire drug industry fascinating. It was only on stepping aside for the arena (retirement is not for me) and finally having the time to think about the 45 years that it hit me just how pointless the entire effort was. Broken does not even describe it because there is nothing there to break. The good news is that the flaws are crystal clear and correctable.

WHERE TO BEGIN

Curing terminal diseases is not a simple task. But it is no more difficult than the other quantum advances humanity has achieved. Those monumental tasks were also accomplished without the enormous capability we now have to process data and collaborate. Unlike the space race, medical research is heavily entrenched in academia and highly regulated. Academia literally plays with science hoping something relevant to disease mitigation will fall out of basic research. The drug industry has no real incentive to cure rather than treat disease, and the regulatory bureaucracy is impeding rather than facilitating cures. The legal establishment and patent system also greatly impede the aggressive approach required to solve the terminal disease crisis. Medical research exists in a world overrun with inefficiencies, and that has to change if we are to succeed in finding the cures that have eluded us. There is no organization, no strategy, and most importantly no accountability for anything that goes on in academia. We trust it to the whims of scientists all running in different directions and chasing career goals that have little to do with cures. So, before getting to the problems with the R&D arena and for the sake of bringing the reader up to speed on the targets to be cured lets dissect each of the disease classes and some of the work that has gone on over the past decades.

CANCER: FIFTY YEARS OF THE SAME TOXIC DRUGS

Cancer strikes fear in all of us. Although the incidence and survival rates have improved over the past 40 years, this progress has largely been due to lifestyle changes, early detection, creative surgical procedures, and greatly improved diagnostic tools. These include new blood tests for tumor-specific antigens and imaging techniques such as diagnostic imaging scans that have permitted earlier intervention. Lifestyle changes such as quitting smoking and eating healthier as well as greater public awareness about the importance of routine checkups have cut cancer rates.

But progress toward a therapy or cure for cancer has been poor at best [1,4]. Many of the same highly toxic chemotherapy drugs that were used in the 1960s are still in use today! The therapeutic principle of these drugs is absolutely draconian. Chemotherapy inhibits the ability of *all* cells, cancerous and normal, to multiply by interfering genetically with cellular replication. The therapeutic principle of these primitive drugs is based on the simple fact that cancer cells multiply faster than normal cells, resulting in their more rapid destruction than normal cells. But both normal and cancer cells are *maimed* by these drugs. Chemotherapy drugs work only by carefully balancing the killing of as many of the cancer cells as possible *before you kill the patient with the drugs*. This is the equivalent of burning down a house to get rid of termites.

The side effects of these drugs are often so bad that patients refuse to take them, despite the consequences. Using these drugs to treat cancers in children or in younger adults is especially risky since these drugs may actually *cause* cancer many years later. Chemotherapy drugs are often used in combinations or "cocktails," which simply combine as many as three to six different generalized cell killers. The advent of the "biotechnology revolution" back in the 1970s was originally touted as the means of developing cures rather than treating symptoms. Unfortunately, 40 years of biotechnology research and billions of dollars spent have not resulted in curing cancer. For example, one of the largest-selling and expensive biotech drugs is Epogen. This is a naturally occurring protein that stimulates the growth of red blood cells. Without these oxygen-carrying cells, all of the body's cells would die. Since chemotherapy drugs kill the cells that produce red blood cells, cancer patients on these drugs become anemic. Epogen was developed to treat that debilitating side effect of these cancer drugs, enabling doctors to administer higher doses of these toxins to patients.

Thus, after billions of dollars in costs to the consumer, not to mention diverting talent from potentially more beneficial endeavors, we are left with an extremely expensive drug that offers at best a marginal clinical benefit. And this story has been repeated many times with many of the biotechnology drugs approved for cancer. Almost every biotech cancer drug must be used in *combination* with the chemotherapy drugs and have significant side effects, so where is the advance with all of this?

Other examples of poor planning and desperation come from two classes of biotech cancer drugs: angiogenesis inhibitors [5] and immune system stimulators [4].The angiogenesis drugs inhibit the formation of new blood vessels, something a tumor needs to grow. Angiogenesis inhibitors are not a cure because it does not eradicate the tumor, but simple reduces its rate of growth. The immune system drugs also have proven to be of limited value [4]. Despite the media fanfare these drugs have received, the results show only marginal improvement in survival time, and even then, these benefits only occur in combination with the old toxic chemotherapy drugs. The progress attributed to the biotechnology revolution clearly has been minimal and an example of poor strategic planning. None of these supposed breakthrough drugs are stand-alone treatments; all are used with the chemotherapy drugs.

Still, in today's environment, a cancer patient's best chance for a cure is often surgery, the same as it was back in the 1960s. One reason why is that surgeons are highly independent and not subject to the same strict regulations as the rest of the medical establishment. Surgeons are relatively free to express their creativity and innovation without FDA approval of their surgical techniques. They are free to innovate and less constrained by academic convention and dogma, something that we will see is a major problem in other medical research arenas. Advances in organ transplantation and improvements in surgical techniques unimagined several decades ago have produced impressive results. Noninvasive surgical procedures have also reduced hospital stays dramatically. Unfortunately, surgical tumor removal is not an option in many cancers or when the initial tumor has already spread to other sites in the body.

A massive effort to cure cancer was launched by the federal government in 1971 with President Richard Nixon's "War on Cancer." One *billion* dollars was allocated back then, but there is precious little to show for it. The money was funneled into an inefficient, nonperforming R&D infrastructure. Perhaps what is needed today is a "Manhattan Project" or NASA-type endeavor in which scientists are pulled together and directed to focus on tangible results with accountability, strategic focus, and timelines. This would certainly address a key problem in the current system, which is the lack of any *global* strategy to find the cure for cancer.

NEURODEGENERATIVE DISEASE: CRIMINAL NEGLIGENCE

The neurodegenerative diseases are the scourge of the last two to 3 decades of life. If cancer does not get you by 60, you have Alzheimer's to look forward to. About 10% of people who reach their 70s can expect to suffer from it or related dementias according to the Alzheimer's Association. This heartbreaking disease destroys the identity and spirit of a person in a painstakingly gradual manner. It destroys patients and families in a different way than cancer since its progression is slow. Nancy Reagan, whose husband President Reagan suffered from Alzheimer's, accurately described it as "The Long Goodbye." As with cancer, there has been much academic research and media attention about research progress, but very little of it has any near-term promise for even controlling the disease, let alone curing it. There is not even a way to avoid a neurodegenerative disease as there is with cancer and cardiovascular disease, and the incidence of the dementias such as Alzheimer's is increasing rapidly.

The basic pathology of Alzheimer's is similar to that of other neurodegenerative diseases, such as Parkinson's and Huntington's disease and amyotrophic lateral sclerosis (ALS). All are equally tragic in what they do to the human body and mind. Different areas of the brain undergo a slow degenerative process and, as a consequence, various physical and cognitive abilities slowly disappear. Neuroscientists have made progress in understanding *how* nerve cells die, but for the past 40 years have made virtually no progress in applying that knowledge to meaningful treatments, more on that key issue of applied versus basic research in the next chapter.

The drugs used to treat Alzheimer's affect the family more than the patient since it makes them feel they are doing something. Cholinesterase inhibitors such as Aricept raise levels of a neurotransmitter in the brain called acetylcholine. This stimulates the remaining functioning neurons that use acetylcholine to communicate with each other. This has a transient effect on the patient's cognition, which unfortunately does not last long, and has no effect on the course of the disease. These drugs only temporarily treat the symptoms.

Neuroscientists now know that nerve cells can be stimulated to death and that almost any injury to the brain can set this process in motion. It has been well characterized in stroke and head injury preclinical animal models, and methods of blocking it have been developed. This phenomenon is called *excitotoxicity* [6], and some researchers are developing drugs to stop this brain damaging process but unfortunately not enough. Alzheimer's disease may be like a very slow form of brain injury [7], possibly initiated by multiple ministrokes, such as a transient ischemic attacks, or TIAs. The resulting excitotoxicity is thought to cause the disease to progress. In fact, one of the newer Alzheimer's disease drugs, Nimanda, blocks excitotoxicity for a certain kind of neurotransmitter called glutamic acid. Although Namenda offers only limited and transient symptom relief, it is a strategy that deserves further exploitation and that has been very slow in coming. Again the strategy issue or lack thereof comes to mind.

For the past 30 years the approach to Alzheimer's treatment has focused primarily on the fatty deposits that are seen at autopsy in Alzheimer's brain tissue [2,3]. These deposits are known as amyloid plaques and resemble the fatty material that serves to insulate normal nerve fibers, or axons. This insulating material is called the myelin sheath, and it is absolutely necessary for nerve cell functioning. For some not-yet-understood reason, this amyloid substance seems to accumulate in the brains of some Alzheimer's patients and in so doing may cause the nerve cells to become tangled and possibly die. I say "possibly" because there has been a major debate in the neuroscience community for the past 30 years as to whether the amyloid plaques seen in brain tissue of Alzheimer's patients *cause* the disease or are an *effect* of it. Despite this debate, the scientific and pharmaceutical communities have spent enormous resources to find ways to destroy these plaques in the hope of stemming the disease's progression largely because of the influence of academic thought leaders.

The quest for clearing the brain of amyloid plaques is only now, after 30 years, beginning to wane. As one might have predicted, treatments that inhibit amyloid plaque formation are turning out to have undesirable effects on normal myelination of nerve cells. In addition, their efficacy in clinical trials is proving to be unremarkable. Why has it taken three decades for researchers to realize this? In that time more progress toward a cure could have been made by pursuing other therapeutic avenues. Quicker progress would certainly come from diversifying resources to many therapeutic approaches and maverick innovation rather than the "flocking" mentality that so often drives basic research [2].

This flocking phenomenon is part of the problem. The basic research community typically moves like a flock of migrating birds. When the leader makes a turn, everyone follows. The difference is that the lead bird knows where he is flying, whereas the thought leader of the scientific pack is often only proposing an untested theory that must be analyzed critically and verified. Amyloid plaques certainly deserved study and exploitation, but not to totally dominate the entire neuroscience effort in the pharmaceutical and biotech industry. Why are there so many relatively useless cholinesterase inhibitors on the market and only one potentially much more useful excitotoxicity blocker approved? Why are not more researchers working on inflammation, autoimmunity, old drugs with potential new uses, natural products, energy metabolism modulators, etc.? All of which have been implicated in the disease, as have a host of other mechanisms. As with cancer, a strategy for finding cures would help a great deal, yet clearly one does not exist.

Parkinson's disease is another potentially terminal neurological disorder in which treatment strategies have remained relatively static for decades. In this disease, neurons degenerate in a very specific brain region that controls movement. The culprit is a collection of nerve cells that use the neurotransmitter dopamine to communicate with each other. In Parkinson's patient's dopamine neurons begin to self-destruct, and when about 80% are gone, movement disorders appear. The treatment that has prevailed since the 1960s is to supplement the dopamine neurotransmitter by giving the patient a precursor of dopamine called L-DOPA, which gets into the brain and generates dopamine. This provides temporary control of the spastic movements, but the effects of the drug diminish with time as the brain's dopamine nerve cells continue to degenerate. Researchers have devised many variations of the L-DOPA therapeutic strategy with a number of new drugs that modulate the dopamine system, but they all employ the same rationale: Treat the symptom and not the cause.

Again, this is a finger-in-the-dike approach. Certainly this now archaic "therapy," unchanged for 40 years, is far from adequate. Why has the explosion in neuroscience research activity had so little impact on its successful application to cure this tragic disease? Few seem to have the innovative spirit to venture very far from the simplistic infatuation with dopamine supplementation. Fewer still have given focused attention to stemming the progression of the disease, or curing it.

Other neurological and neuromuscular diseases that suffer from the same dismal failures in biomedical research include Huntington's disease, ALS, and muscular dystrophy (MD). How many MD telethons will it take before any meaningful improvement in the prognosis for MD patients is forthcoming or before the public realizes that something is not working? Are we going about finding a cure in the proper manner? Is it a question of money or execution? Certainly there is no lack of money given the $32 billion annual NIH budget and the numerous universities, charities, and foundations, so something must not be working in our approach. Meanwhile, people continue to die. We must turn over more stones and use *multiple creative insights* rather than the regimented narrowly focused approach we currently practice.

CARDIOVASCULAR DISEASE: TREATING THE SYMPTOMS

Cardiovascular disease, including heart attacks and strokes, is the number one killer in the United States. It's not hard to understand why if you understand how our cardiovascular system functions. It's a living plumbing system with the life-sustaining job of delivering oxygen and other essential substances to every cell in the body. Interrupt that for several minutes, and life slips away. Problems with our human plumbing are usually caused by clogs, i.e., blood clots. The drug industry has made significant strides in treating the symptoms but not the causes. We are better at dissolving blood clots acutely during heart attacks and strokes when the plumbing gets totally clogged in one spot, and this has reduced deaths from acute heart attacks and strokes. Most of the progress, however, has come from changes in lifestyle, nutrition and diagnostic procedures, which have delayed the deaths from heart attacks and strokes. There is, however, no cure for either heart attacks or strokes. There are scores of drugs to reduce blood pressure or lower cholesterol, which is thought to reduce the risk of damage to the heart and blood vessels, but no one has figured out how to keep the plumbing from clogging up in the first place or how to purge it once it has. Stents and bypass surgery can repair small parts of the circulatory system, but the cure would be to flush out the entire system and that has not happened, why not? If there were a strategy in place rather than just waiting for scientists to decide to do it, this would not be the case.

Surgeons, in partnership with interventional medical device manufacturers, have made great progress in mechanically cleaning out small sections of arteries. Their skill and creativity shine in such procedures as coronary artery bypass grafting surgery and angioplasty. Is it coincidental that the surgeons who have made these strides are not extensively regulated by the FDA? The clogging continues, as does the enormous loss of life. Clearly we need to find a way to keep the plumbing open and to treat the cause rather than the effect. The majority of the effort in this area continues to be generating more me-too drugs to treat the effects. That needs to change.

TIME TO GET SERIOUS

Why is not the public outraged about the lack of any real progress in curing the terminal diseases? Certainly we are paying a fortune for this monumental failure in both dollars and lives. The Ivy League universities and the NIH are rich with capital. Why are we getting so little in return, and is that state of affairs inevitable? The dearth of treatment breakthroughs is being overshadowed by all the fuss the media makes about "scientific breakthroughs." We are constantly bombarded by media reports about various highly technical aspects of what medical scientists are unfolding in the laboratory. But it never seems to lead to anything relevant regarding cures or paradigm shifts in disease therapy. These small advances lull us into thinking there is light at the end of the tunnel. Just how long is this tunnel, and how much more media hype regarding breakthroughs must be tolerated before we lose our patience with this fiasco that is medical R&D?

Our culture's love affair with science has conditioned us to absolve its practitioners of accountability. We are repeatedly seduced with science, even if we do not seem to understand it well. Gene therapy, stem cells, mapping the human genome, all very attractive but after decades no cures have emerged. It is always very preliminary, extensively hyped and used to paint a picture that cures will be forthcoming. Strategies and timelines to a cure from this science are rarely discussed. We leave this all to scientists who are, by definition, more interested in the phenomena they discover rather than its application. It is time to demand accountability and restructure the entire medical research establishment. We can no longer tolerate academic games that play with potential cures.

Some examples of the hyping of science include the biotechnology craze, gene therapy, the human genome project, and most recently, the stem cell hysteria that swept Hollywood, the media, and state legislatures. Billions of dollars have been appropriated to chase these science excursions with little critical examination of their potential value. The California stem cell imitative of 2004, which spent almost 3 billion dollars but has produced almost nothing, is a good example [8]. The only certain result of these efforts has been to further bloat an uncoordinated unaccountable medical research infrastructure. This does little to achieve the results we so desperately need. The record is clear: We have failed at the business of curing terminal disease. A good analogy for our current situation is an orchestra without a conductor or sheet music. Everyone is playing their own tune, and the music never happens. This can and must be changed.

REFERENCES

[1] S. Begley, M. Carmichael, Desperately seeking cures, Newsweek Magazine (May 14, 2010). Retrieved from: www.cureswithinreach.org/newsroom/media-coverage/137-desperately-seeking-cures.

[2] S. Begley, Is Alzheimer's field blocking research into other causes? The Wall Street Journal (April 9, 2004). Retrieved from: www.wsj.com/articles/SB108145279348578177.

[3] S. Begley, Scientists' world-wide battle a narrow view of Alzheimer's causes, The Wall Street Journal (April 16, 2004). Retrieved from: www.wsj.com/articles/SB108206188684384119.

[4] M. Konner, Watch the hype: cancer treatment still has far to go. Two decades' achievements are modest, Wall Street Journal (March 17, 2016). Retrieved from: http://www.wsj.com/articles/watch-the-hype-cancer-treatment-still-has-far-to-go-1458231884.

[5] Angiogenesis Inhibitors, National Cancer Institute, November 19, 2016. Online Retrieved from: https://www.cancer.gov/about-cancer/treatment/types/immunotherapy/angiogenesis-inhibitors-fact-sheet.

[6] M.R. Hynd, H.L. Scott, P.R. Dodd, Glutamate-mediated excitotoxicity and neurodegeneration in Alzheimer's disease, Neurochemistry International 45 (5) (October 2004) 583–595.

[7] W. Ong, K. Tanaka, G.S. Dawe, L.M. Ittner, A.A. Farooqui, Slow excitotoxicity in Alzheimer's disease, Journal of Alzheimer's Disease 35 (February 12, 2013) 643–668.

[8] B. Fikes, California stem cell agency charts course for future, The San Diego Union Tribune (December 18, 2015) A1–A11.

Chapter 2

Academia Is Marching to the Wrong Drummer

Chapter Outline

The great majority of the academic medical research establishment resides in universities, the National Institutes of Health (NIH), and many other private institutes such as the Salk Institute, Scripps, and hundreds more. It is all funded by the NIH and various foundations such as the Hughes Foundation, the National Science Foundation, and many others. Academia is huge, uncoordinated, and revered by society in general; a reverence that has managed to escape from the checks and balance of other endeavors.

SCIENCE FOR ITS OWN SAKE

The biomedical academic research infrastructure is focused on goals that have little to do with cures. These include academic freedom, teaching, tenure, and research awards. There are thousands of PhD professionals working in university labs, as well as scores of public and private biomedical research institutions. Many pharmaceutical and biotech companies leverage off academic basic research in an effort to translate it into products. Trillions of dollars each year are spent to support this huge empire, and the only way to describe it is chaotic.

With all due respect, having been a part of it for 15 years of my career the best analogy that I can think of to describe it is the Keystone Cops. Everyone is busy and excited but totally disorganized and uncoordinated with regard to each other. There is no strategic plan or attempt to coordinate what is going on at one university with another. Redundancy is rampant, accountability is nonexistent, project managers are nowhere to be found, and the bottom line solutions

A Roadmap for Curing Cancer, Alzheimer's, and Cardiovascular Disease.
http://dx.doi.org/10.1016/B978-0-12-812796-4.00002-7

19

for real-life problems that emerge from this chaos are few and far between. Basic researchers are literally playing with science, chasing notoriety from their peers, and focused on tenure, all of which are contrary to what needs to happen to end the cure crisis. No one tells them what to focus on; no one audits their data or dares to challenge their academic freedom. They are their own boss chasing science under the veil of academic freedom. The application of that science is at best a secondary down the road goal that gets mentioned in grant applications but rarely happens.

Most academic life scientists derive their rewards and motivation from pushing the boundaries of knowledge to generate new hypotheses for experimental testing. They are not able or inclined to take their data, or more importantly, someone else's, and apply it to real-world problems. This diversion of mission from application to advancing science has cost many lives since much of the science being sought is not relevant to disease or too virgin to be of any value for decades.

The academician's training does not include what every patent application must have: *reducing your innovation to practice*. They are trained to generate new hypotheses' based on their data and move on, not to translate their last discovery into a product. To make matters worse, applied research is repulsive to the academic mentality and synonymous with a lower form of science. The end result is an accumulation of data, which gets published and then, due to the patent codes, becomes useless from a commercial perspective because *no one owns it* and published data cannot be patented. We are therefore literally drowning in science, and it is not being applied because the scientist is motivated to publish data rather than patent and in so doing condemn it to commercial obscurity. The patent system has not responded to this reality by changing the rules on patenting published data, something we will discuss later. Publishing builds the academic scientist's career and gets him or her recognition, tenure, grant funding and awards, its publish or perish. Again, I played this game rather successfully in my early career, and I can attest to the lure and excitement of it but I soon became impatient with where it was leading. The bottom line is that the medical science academic is locked into a system of science worship that inhibits applied research.

CHAOTIC POOR-QUALITY SCIENCE

The life of an academic biomedical scientist revolves around getting research grants. While relevance to health and disease-related studies is a criterion for getting grants, it generally just has to be mentioned by the grantee as a down the road purpose of the research and is not the primary goal with cure-related milestones and accountability for reaching them or trying another strategy. Having sat on many grant review panels it is the quality of the science and the novelty of the work proposed, coupled with the credentials of the principal investigator (PI), that determine who gets funded. There is little consideration given in the granting process to a strategy of how to achieve a given goal, such as curing a disease. It all hinges on the novelty of the science, not the likelihood of it producing a cure or how it relates to or coordinates with other work being done elsewhere to achieve that goal. There is no big picture but rather the science is evaluated in a vacuum, not as part of a strategic vision. Big problems do not get solved that way.

The biggest problem in academic research is that there is no data quality control criterion in the granting process. For example, research in industry is done under good laboratory practice or GLP. The Food and Drug Administration (FDA) mandates this, and the GLP guidelines are very strict. They insure that no investigator bias is possible and that all the conditions under which the data were generated can be reproduced. Academia and granting agencies do not follow these GLP practices and this leads to many irreproducible results and false starts in academic science. Incorrect theories remain alive and lead to a lot of wasted time. Consequently today's breakthrough can become tomorrow's irreproducible result or failed clinical trial, and we never hear about it again [1,2]. This is a huge problem since many start-up companies are formed on this basis; a reason so many start-up companies fail.

The granting process is also extravagant with its capital and is a huge source of income for the universities and institutes that employ the PI. It is called overhead or indirect costs. The university generally gets anywhere from 30% to 100% of the grant amount for overhead. Most of the faculty whom I have spoken to have no idea what expense this money covers, other than to keep the lights on. Also to make matters worse the grant covers a large portion of the salaries of the university investigators. Most science faculty receives their entire salary and benefits from grants. One wonders where student tuition goes and how the university establishes a cost basis for the tuition they charge. The bottom line is that the granting process is flawed in its waste of money since a large percentage goes to the university and not research [3–6]. Essentially half or more of all academic grant money does not go toward research, and the sponsoring institution also gets the bonus of having the professor and his or her staff free of charge, benefits included.

UNIVERSITIES: WHERE IS THE ACCOUNTABILITY?

Clearly there is a lucrative and occult business at work when it comes to universities and their cash flow. The rationale for how grant money gets us any closer to cures is nowhere to be found. The university employs the faculty to teach students for which they charge tuition supposedly to pay faculty and staff and support its infrastructure. The faculty, however, must get grants awarded by peer review (their colleagues from other universities sit on these review panels, perhaps a conflict of interest?), which pay their salaries so the university now gets a free ride as well as a huge chunk of overhead money as a bonus.

Then there is the endowment fund. This derives from alumni contributions, the endless letters and phone calls from the alumni fund that persists until, and often after, the alumnus perishes, as well as numerous other sources. Most universities have hundreds of millions of dollars in their endowment funds and tens of billions in the Ivy League. What a racquet; an uncoordinated, unaccountable system that controls huge sums of research capital and is not answerable to anyone. All these, while tuitions increase at multiples of the inflation index, graduates suffer the burden of enormous loan repayments now exceeding a trillion dollars and taxpayers footing the bill. Does this make sense in the context of the mission at hand?

Our infatuation with academic basic science has led to multiple potential breakthroughs but no cures. The typical scenario in medical research is that scientists will discover a biological process, largely at random, and then ask whether it might have a role in various diseases. While this strategy can generate new disease therapies, it is, at best, haphazard and has yet to work. It often leads to a "begging the question" thought process where the investigator designs experiments to demonstrate that his or her breakthrough finding is relevant to a particular disease while ignoring the downsides. This is science in reverse, and such a shoehorn approach often leads to dead ends as has been the case with biotechnology, gene therapy, and stem cells as we shall see later.

When virgin science makes a connection with a disease, industry gets interested and proceeds to develop drugs or treatments based on it. This generally takes the form of venture capitalist–backed start-up companies (a waste of money in most cases as we will see later) since the larger pharmaceutical companies are too sophisticated to jump in at this point. The media begins to trumpet the new discovery and its potential for curing a disease, and the frenzy begins. Multiple, venture-backed companies are formed only to find out years later that the strategy failed. Anyone who has invested in these companies knows the drill and has felt the financial pain when these ventures go belly up. Lawyers are the only beneficiaries in this scenario. The result is many years of wasted effort, failed commercial enterprises, and, more important, the diverting of resources that could have been more effectively used if any measure of strategic planning was involved. What is needed is a frontal assault on a specific disease, not the random, often biased, application of preliminary unproven science.

The path to commercialization that preliminary science findings take is highly inefficient and totally lacking in strategy. Few nonscientists understand the technologies involved, and therefore investors have to rely on the academic scientists themselves as advisors thus creating an enormous conflict of interest. The record of venture capital–backed biotech start-up companies over the past several decades has consequently been horrible when judged by the development of disease cures. To make it even worse, most of these venture-backed companies are plagued by the same academic mentality that has hampered applied research innovation in academia. Often the venture capitalists are duped by the academic thought leader advisors they routinely consult before funding such companies. The media, in its hunger for news, facilitates the entire mindless process by overstating the potential for virgin science as the potential cure for cancer or some other disease, and it is off to the races again chasing yet another rainbow.

Biotechnology in general is a great example of how academic research was exaggerated. It was sold to the public and the investment community as a breakthrough and is an example of how science and technology can be hyped and sold to an unsuspecting investment community. As we shall discuss in detail in Chapter 4, biotechnology is simply a technical means for making proteins outside the living cell in the laboratory. Most of the major biotech products are simply proteins whose functions were previously known but are now easier to produce outside the body using biotechnology. Therefore, the entire scope of biotechnology is simply a way of making something more efficiently and not a new understanding of the biology of disease. Biotech drugs such as TPA and human insulin and various antibodies have been around long before biotechnology came on the scene. Yet, these constitute a major portion of the "breakthrough" biotech products on the market today. The disappointing truth is that the biotech industry has been ineffective as an innovator and has failed to be the promised agent of treating the cause rather than the symptoms. This 30-year excursion has consumed hundreds of billions of dollars, produced mainly failed companies, and a handful of high-priced products that have provided incremental medical advances. Sadly, the same is true for gene therapy, stem cells, and the human genome project. All are great science and that science should be continued but not as an alternative for established science whose innovative exploitation could produce near-term treatment breakthroughs.

The public and the investing community have supported all of this in the misguided hope that disease cures would automatically result from these technologies. The "if it is new, it must be better" mentality is clearly wrong for near-term cures and serves academia better than it does society because it means more grants can be obtained to chase it. Science sells well, especially when it emanates from prestigious places, but unfortunately, that has served to cloud the reality that disease cures are not happening. Less fanfare regarding the science and more focus on the results are needed. But the universities have benefited from this misguided diversion.

TECHNOLOGY TRANSFER, WHY BOTHER?

Universities realize that faculty professors are not entrepreneurs and they are trying to transfer faculty-based technology into commercial applications by making them available for licensing. Most have established "technology transfer" centers on campus that attempt to bridge the science to product transition. These centers are well intended but have had limited success since what they are selling is often poorly vetted preliminary data with limited patent protection and no quality control. The faculty spends little time exploiting the technology transfer resource, and universities are just plain bad at commercialization. The patents they generate are poor due to inadequate resources allocated to their patent filings, which limits their value to industry. For example, many times only provisional patents are filed, which only provide 1 year of coverage, and when a full patent is filed, it is not always followed by filing patents in the other major territories such as Europe and Japan due to the expense involved. Universities also lack the skills to negotiate workable licensing agreements with business entrepreneurs, and they often have State or Federal rules and constraints on licensing that complicate working with the business sector in general.

Having dealt with several top tier universities in search of novel approaches to disease treatments, I can attest to the fact that it is a frustrating experience. The initial meeting goes well but that is where it ends. State universities are especially difficult to deal with since they often have procedures and mandated restrictions in dealing with the evil folks in industry that are very hard to navigate. The university is typically not willing to take *any* risk and often requires too much up-front in the form of licensing fees from the entrepreneur for it to be feasible. This impedes the entrepreneur's ability to move forward. The major problem is that the data being offered for license are poorly controlled from a quality perspective because the faculty member who produced it was motivated by publication and not commercialization and the data quality required for each goal is entirely different.

THE COLLABORATION MYTH

Scientists often speak about the value of collaboration. This is probably one of the biggest delusions in the biomedical research establishment. The nightmare of virtually every basic scientist is the thought of being "scooped" on an important research finding. This simply means that another researcher publishes the same data that you are working on before you do. One of the commandments of academic science is "Thou shall not be scooped." I once left an international medical conference several days early to write up a paper after seeing a presentation that was similar to my work. This quest to be first often fosters extreme secrecy and rushed publication of unvetted data rather than the open collaboration and sharing of data that drives science progress. Collaboration in biomedical science does occur but it often takes the form of recruiting a colleague who knows how to perform a technique and asking him or her to analyze some samples for you using that technique in exchange for coauthorship on a paper. It rarely has anything to do with the pooling of brain power to solve a problem or teaming up with a fellow scientist to facilitate the transfer of their technologies into a real-world product. The public deludes themselves into thinking that this happens at scientific meetings. The reality is that everything presented at these meetings is published in abstract form weeks before the meeting and most scientists attend for pure networking purposes and to get ideas rather than share their own. To make things worse the abstract is a publication so forget about a patent, which is absolutely required to commercialize the technology.

There is rarely any serious brainstorming that goes on between rival laboratories when it comes to designing research strategies to answer fundamental questions. That extremely important process often occurs in secret to avoid losing credit for discovering a potential important advance. This was blatantly demonstrated by the dispute between French and American investigators over who should get credit for discovering the AIDS virus. That debate, which got quite ugly for a while, involved the NIH and the Pasteur Institute accusing the other of scientific misconduct regarding the Nobel Prize for medicine in 2008 and illustrated how important it is to the basic scientist that credit for his or her discoveries be properly recorded [7]. One can only wonder where we would be today if that competitive energy was directed at true collaboration to solve the fundamental problems of terminal diseases.

THOUGHT LEADERS: BLESSING OR CURSE?

Thought leaders are bright, opportunistic academic scientists, often with an Ivy League lineage and a postdoctoral stint working under an older thought leader or Nobel Laureate. They are proponents of the hot and trendy science areas that attract the majority of the brightest scientists. They have a great deal of grant support and large laboratories with many postdoctoral fellows working on the thought leaders' pet project. Thought leaders are on all the editorial boards of the most prestigious peer review journals. Many have Nobel prizes and other impressive awards. While they are bright and impressively credentialed, they do not have a corner on the innovation market. And as we shall see, innovation always trumps intellect when it comes to the quantum leaps required to cure disease. Thought leaders, however, wield great power in the research establishment. The media follows them, and investment capital tends to flow where they say it should.

This infatuation with thought leaders appears to make sense on the surface; after all, they are the experts in their given areas. It does, however, breed a "flocking syndrome" that can be absolutely disastrous since it limits the scope of ideas that get tested and innovation in general. Typically, multiple companies are formed to test a single thought leader hypothesis. Feeding frenzies form if the hypothesis is from a Nobel Laureate. The range of creative and innovative activity that would prevail otherwise is therefore limited.

Thought leaders tend to become heads of institutes and research foundations or university presidents, deans, and department chairpersons, etc., where they exert a great deal of influence on other more junior scientists. This influence again serves to stifle innovation and transform progress into a spiritless form of "follow the leader." The media makes them into prophets of "the next science breakthrough" or calls them the "science rock stars" when in reality, they do not have a corner on innovation. Unfortunately, once the box they create is formed, it is very difficult for the innovators to scale its walls. Science is about simplifying the complicated and seeing the world from many angles. Unfortunately our thought leader oriented system works against that mentality leads to regimentation and failure especially when the problem to be solved has escaped resolution for decades.

Tenure is another enemy of innovation. The simple admission that one has now reached a point in his or her career where he or she is essentially beyond reproach is a license for stagnation. One can only speculate on how many worthy ideas have been thwarted by the granting of tenure at universities or receiving a career changing award. As we will see in later chapters, cures often come from unexpected places and maverick voices and less often thought leaders basking in the comfort of tenure and busy messaging their fame.

The human genome project is a good illustration of how a group of thought leader molecular geneticists convinced the powers that be to spend billions of dollars on the mapping of the entire human genome. The rationale was that this information would provide many new targets for disease treatments. The media, in its typical clueless enthusiasm, fanned the flames and the public was immediately on board. The gene therapy folks claimed that the availability of new targets (genes) would help them find cures. However, one enormous fact was overlooked and that is: The gene therapy folks already had more targets than they could handle *before* the human genome project was ever conceived. There are scores of diseases where a specific gene defect is known (right down to the specific amino acid in the protein in some cases), as the cause and has been for decades, but cures for these diseases have not resulted. These include sickle cell anemia, Huntington disease, dwarfism, Down syndrome, and a host of others.

There is a very thick book dedicated to these diseases titled, *The Metabolic Basis of Inherited Diseases*. For the past 40 years, not one single gene-linked disease in this book has been conquered by gene therapy. So one might ask: Why spend billions of dollars to map the entire genome when the numerous targets we have had for decades remain unexploited and unsolved? Is this a useful goal or is it a purely science and media-driven project to serve thought leader career goals and egos? This is not to say that having a human genetic map is useless, but it is probably not going to save many lives any time soon so why not put the money somewhere that would. Also, messing with the genome is certain to be a very risky strategy with side effects that could be scary and irreversible. A priority-driven strategic plan for curing disease would not have included such an expensive luxury as a top priority. Three thousand million dollar (total cost) grants could have been funded with that three billion dollars or 4500 grants if indirect costs were brought under control.

Stem cell therapy is another example of nonexistent strategic planning, and exuberant thought leader-driven media abetted science worship. Here we have a preliminary set of findings that focus on a precursor type of cell that can develop into any number of specific cell types, such as a nerve cell or an insulin-producing cell. The rationale is that putting these cells into an Alzheimer's brain or a type I diabetic pancreas will miraculously cure them. The thesis is seductive, but the reality is most likely very far off and in need of much critical scrutiny before we waste billions of dollars on premature crusades. Nevertheless, California immediately floated 3 billion dollars in bonds 12 years ago to fund stem cell research, which no one quite knew how to spend [8]. Actions such as these plants hope and keep the public satisfied that something is being done and that the science gurus are excited about it. The California effort is now almost out of money and still trying to salvage something out of it. This again demonstrates the ill-thought-out hysteria that thought leaders can, with the help of the media, sell to the public. The 3 billion dollars spent on such a pie in the sky research project that no one knew quite how to manage could have funded scores of innovative companies, again, if a strategy existed.

A good analogy to the stem cell research situation is to imagine a society that has just discovered concrete and now plans to build the Empire State building. Stem cells may be the concrete, but what about all the other elements that are required to build a skyscraper? Is it rational to think that an elegant and infinitely complex structure like the human brain will just magically accommodate stem cells dumped into it and proceed to assemble them in the precise way required to cure itself? It is not just a matter of having the nerve cells; they also need to be in the right place at the right time, with other supporting elements; just as it takes steel, wires, and pipes, along with concrete, all applied in the *correct* sequence, to build a skyscraper. Inserting stem cells into a mature organ will likely have ramifications that may be worse than the disease being treated. How do you keep them from overgrowing and becoming rouge cells? How do you deal with the cancerous tumors observed in stem cell laboratory experiments [9,10]? Cancer cells and stem cells share many common characteristics. How do you deal with the underlying disease process that caused the cells to malfunction in the first place? Doesn't it make sense to answer these questions before we rush head long down this path to developing disease therapeutics with a technology we do not understand?

The public, and the funding establishment, needs to understand how much pure faith is involved in assigning a therapeutic or curative role to biotechnology, gene therapy, the human genome project, stem cells, and a host of other thought leader propagated theories. The academic passion for all these basic research concepts is indeed appropriate, but presenting it to the public as a potential wonder cure for terminal diseases is irresponsible and a waste of resources. If there were a strategy for curing terminal disease, stem cells would be assigned a longer-term project whose feasibility remained to be determined rather than a bandwagon that had to be jumped on.

THE NATIONAL INSTITUTES OF HEALTH: WHERE IS THE BEEF?

The NIH is a government-funded institution focused on health research, with separate institutes devoted to specific groups of diseases. The mission statement, however, is totally silent on curing any disease let alone the terminal diseases. I spent 13 years there and had a ball doing whatever I pleased, publishing over 240 research papers, editing 4 books, and founding the *Journal of Molecular Neuroscience*. All these got me tenure in record time, awards, and invitations to speak globally, but none of it was related to curing anything. If you filled the journals and developed an international reputation, that was all that mattered. These success metrics are the root of academia's problems as it relates to solving urgent problems and until they are dealt with will continue to impede progress.

The NIH conducts intramural research at its own laboratories and provides the funding for much of the extramural research that takes place at universities and research institutes throughout the United States. It controls the allocation of tens of billions of dollars per year to academia, all of which is focused in theory on health-related research. The NIH has produced enormous quantities of basic research data, and it "attempts" to test new therapies on patients at the clinical research hospital in Bethesda. The reason for the quotation marks will become apparent later. While these all sound exciting in principle, cancer still kills 600,000 people each year, virtually no progress has been made on curing Alzheimer's and cardiovascular disease is still the number 1 cause of death taking over 800,000 lives each year.

The NIH was a noble gesture that sprang from the realization that academic medical science should be focused on health. It provides its in-house scientists with funding for their research, freeing them from having to write grants and is basically an academic rather than the goal-oriented timeline-driven organization required to cure disease. It is essentially a federal university rather than a center for applied research with the major incentives being academic freedom, publishing, tenure, being invited to scientific meetings, awards, and peer recognition.

The separate institutes at the NIH are split up into laboratories or branches and the laboratories into sections. Each laboratory or branch chief is analogous to a university department chairman and each institute director to a dean. It was conceived as a university and basically operates as one. The leaders are research professionals with no training or experience in transitioning research into products or in developing implementation strategies to accomplish that. Furthermore, there is no concerted effort to direct the scientific staff toward specific disease-related goals with milestones, timelines, and accountability. The mentality is essentially the same as that at the university; generate new knowledge and attempt to cherry-pick it for its therapy potential rather than making science a tool to treat disease directly. There is very little collaboration within laboratories and the mindset was one of completion, academic freedom, and getting tenure.

What is clear to me now after spending over 30 years in the industry is that the NIH is not designed to cure anything but rather as a shrine to the health sciences that panders to each special interest group in the health arena. There are now 27 institutes at the NIH, while there are only three terminal disease classes that are responsible for over 95% of all the fatalities! This in itself speaks to a total lack of priorities and focus. Abundant research has emerged from the NIH, but the results speak for themselves. The problems at the NIH are the same as the rest of academia with the differences being the lack of classrooms and freedom from writing grants constantly to support your research. It needs a major overhaul to focus and get serious about the tragedy it pays lip service to. I have great respect for the talent within its walls, but they are wandering in their own universe of basic science and need to redefine the mission statement to focus on the conquest of the major killers; more on the specifics of what needs to be done to accomplish that in Chapter 7.

STRATEGY, ACCOUNTABILITY, AND URGENCY: THE MISSING LINKS

The Manhattan Project and the Apollo Project are examples of how goal-directed science can achieve miraculous results rapidly. In both cases, a combination of scientific talent, strict management, and a close collaboration with private industry all came together to create miracles thought to be impossible and did it ahead of schedule. There was no redundancy of effort or infatuation with science for its own sake or the host of other academic diversions that cloud progress. The aerospace and physics scientists were not hostile to the aerospace industries but were rather partners working together on a common goal. This is not the case with medical scientists and the drug industry. There were risks involved that were accepted, but what they did was not strangled by regulation and law suits. The science served the goal, rather than it being an end in itself. If academic scientists, universities, research institutes, and charities were asked to generate a moon landing or harness energy from the atom or conquer cyberspace, it would still be in progress. This single-minded purpose coupled with organization, strategy, accountability, and urgency have yet to be applied to curing terminal disease and until they are we will continue to crawl through this crisis.

What has kept us all quiet and in awe of an academic research juggernaut that has produced little in the past 50 years? Some reasons are quite obvious. First the public is easily fooled when it comes to medical science. They do not understand the basics of what these diseases are and that needs to change; hopefully this book will do that. Another reason is that the media has effectively pacified us with their constant exaggeration of basic science breakthroughs that promise cures. The science establishment and the institutions that house it have become sacred cows beyond accountability. We continue to give to universities, foundations, charities, and research institutions, yet all we see is impressive infrastructure and random data, not results.

Another reason the public is brainwashed into submission is the apparent progress relating to cancer incidence and survival. Survival rates have improved for some types of cancer. But this is due largely to early detection, which is the result of research breakthroughs in the field of physics (CT, PET, and MRI imaging techniques) and not biomedical research, as we discussed earlier. Early detection is also a result of improved awareness by the public itself, leading to more frequent screenings for cancer. The reality is that the public has done more for itself by lifestyle changes (smoking cessation, diet, and exercise) and getting more frequent checkups than has biomedical research with all of its new science. The same can be said for cardiovascular disease. Thirty years ago you had to search for a health club; today they are as common as supermarkets and everyone monitors their fat intake and floods themselves with free radical scavengers that are thought to block disease processes such as inflammation, which plays a role in all of the terminal diseases.

The academic medical research complex has to be totally rethought with far less focus on random uncoordinated, unaccountable science, and much more on urgency and managed, milestone-driven progress metrics. This is not about Nobel Prizes and building universities with endowments that match the treasuries of some countries; it is about saving millions of innocent lives and that reality seems to have been forgotten. We need to get serious about how to exploit medical science rather than play with it. We need much more innovative applied science to achieve near-term progress and a more rational approach to new initiatives that will take decades to bear fruit, if at all. The time wasted to date is a tragedy that if permitted to persist will continue to take several millions of lives each year. Our scientists have failed us as we have failed ourselves by not dealing with that. They are talented and dedicated to science, but we have not provided them with the generals and the strategy to channel that talent toward solving the cure crisis. I am proud to be one of them. Nothing gives me more pleasure than talking science, but there comes a time when that has to be put in perspective and we must sacrifice academic freedom for the needs of society. Science for its own sake is not the only problem as we will see in the following chapters.

REFERENCES

[1] G. Naik, Mistakes in scientific studies surge, The Wall Street Journal (August 10, 2011) A1–A12.
[2] A. Marcus, Lab mistakes hobble cancer studies by scientists slow to take remedies, The Wall Street Journal (April 21, 2012) A1–A12.
[3] F. Mussano, R.V. Iosue, Colleges need a business productivity audit, The Wall Street Journal (December 28, 2014). Retrieved from: www.wsj.com/articles/frank-mussano-and-robert-v-iosue-colleges-need-a-business-productivity-audit-1419810853.
[4] A. Cave, Taking a hard look at University research, Stanford Social Innovation Review (October 20, 2014). Retrieved from: https://ssir.org/articles/entry/taking_a_hard_look_at_university_research.
[5] H. Ledford, Indirect costs: keeping the lights on, Nature (November 9, 2014). Retrieved from: http://www.nature.com/news/indirect-costs-keeping-the-lights-on-1.16376.
[6] V. Callier, Overspending on overhead, The Scientist (February 1, 2015). Retrieved from: http://www.the-scientist.com/?articles.view/articleNo/41962/title/Overspending-on-Overhead/.

[7] R. Bazell, Dispute behind Nobel Prize for HIV Research, NBC News, October 6, 2008. Retrieved from: http://www.nbcnews.com/id/27049812/ns/health-second_opinion/t/dispute-behind-nobel-prize-hiv-research/#.WDEpXI8zVdg.

[8] B. Fikes, California stem cell agency charts course for future, The San Diego Union Tribune (December 18, 2015) A1–A11.

[9] G. Prindull, Hypothesis: cell plasticity, linking embryonal stem cells to adult stem cell reservoirs and metastatic cancer cells, Experimental Hematology 33 (2005) 738–746.

[10] P.C. Herman, S.L. Huber, T. Herrler, A. Alcher, J.W. Ellwart, M. Guba, C.J. Bruns, C. Heeschen, Distinct populations of cancer stem cells determine tumor growth and metastatic activity in human pancreatic cancer, Cell Stem Cell 1 (2007) 313–323.

Chapter 3

Mavericks Versus the Establishment

MEDICAL RESEARCH 101

There are two motivations for doing science. Basic science is the pursuit of knowledge for its own sake, while applied science is the innovative utilization of new and old science to solve a problem or create a product. Applied science is using science as a tool for a purpose, while basic science is wandering in the unknown and pushing the boundaries of knowledge. The former is designed to find near-term solutions, is pragmatic, and is not very romantic. The later may provide long-term solutions, is not necessarily designed for that, but rather just to know, and is very romantic. Academia focuses on basic science, and industry exploits applied science. Basic scientists are the frontiers men and women who push the boundaries, win the prestigious awards, get tenure, and become department chairpersons. They are the journal editors, the grant reviewers, and the thought leaders of the academic medical establishment and are generally considered beyond reproach and immune from criticism. The media calls them the rock stars of science and makes them famous. They cannot quite grasp why they are rock stars, but they pin their hopes on them and assume cures are around the corner because of them. Science is in many ways the new religion built around knowledge.

Applied science is the stepchild of basic science and is viewed as a less pure intellectual endeavor by academicians. It simply takes what is known and applies it to a problem. The problem is that there is far too little applied medical science being done in academia. Academics shun applied science and prefer to push the boundaries of science and rush to publish their data before it is even replicated by someone else, let alone patented. Basic science suits the goal of publication since the questions to be answered are endless, and each new science finding generates multiple hypotheses to be tested and published, which is what academia thrives on. The glut of unexploited science in the literature is the result of this rush to publish, and it constitutes a major impediment to the practice of applied research. This begs the question: Why not file a patent first? Because scientists do not get the peer recognition that comes from publishing and it costs money to get the legal assistance required for patent writing and the numerous office actions to get the patent allowed. In addition, universities provide minimal support for patents, and granting sources generally do not require or encourage it. The academic inventor is left with little choice and takes the publication route rather than file the patent.

ESTABLISHED SCIENCE IS BEING WASTED

There must be more applied research in academia. Academics need to focus on cures by using whatever is out there rather than building their own knowledge estates that may not be relevant to such application. Science needs to take a back seat and return to the status of a means to an end and not the end in itself. We need more potential paths to cures, not just blind faith in virgin technologies that at the very best will take decades to produce any fruit. How did this happen?

Applied research was shunned back in the late 1960s when universities responded to the antiestablishment mentality forced on them by students and shifted their mission away from commercialization. Institutions, again responding to student uprisings, began to shun collaborations with industry as a way of insuring that the knowledge they generated was pure and free from exploitation. I saw this happened when I was at the National Institutes of Health (NIH). During my 12-year tenure there, it became increasingly difficult to collaborate with industry and that trend has continued. This explosion of antiestablishment sentiment during the 1960s and 1970s cast all of the business community in a negative light, and along with it the notion that intellectual pursuits should be free from commerce or profit. The intellectual community has since shunned applied research as irrelevant to their quest for new knowledge and the exercise of their academic freedom. Commercialization was increasingly considered a cop-out since it implied profit and that had a bad connotation in academia. Applied research necessitates a commercial partner and that became much harder to do.

Industry is now forced to focus only on very new, preliminary, unproven technologies that are patented for its drug development efforts rather than the broad spectrum of approaches that are available in the literature but lost to commercialization. New technologies are by definition the most risky, so the observed failures and false starts were to be expected, but what other alternative is there if these are the only technologies that can be patented? Until we lift the veil of patentability and the restrictions on academia–industry collaborations, this enormous storehouse of knowledge in the literature will be wasted and progress will be far slower than it could be. In Chapter 10 we will talk about how the patent laws can be changed to open up published science to patents but for now it is sufficient to say that academic medical scientists need to exploit existing science with focus on the disease target and not just their own specific interests.

MAVERICK SCIENCE, OUR BEST HOPE?

The typical career path taken by academic scientists is to obtain a PhD or MD, and then pursue a postdoctoral fellowship. The postdoctoral fellowship serves as a consummation of the PhD much the same as the internship and residency consummates the MD. Freshly minted PhDs seek out thought leaders as their postdoctoral mentors and typically spend 2–3 years working in their laboratories on projects of interest to the thought leader. The postdoctoral mentor often plays a pivotal role in getting the matriculating scientist his or her first full position, as well as steering them along on their career paths. This was the case with me. My postdoctoral mentor's department chairman played a big role in getting me my NIH position since he was formerly an NIH investigator with excellent credentials and connections at the NIH.

The postdoctoral fellow often becomes a disciple of the thought leader. In fact, those fellows who follow their mentor's area of work the closest can progress upward the fastest. Grants are generally easier to get as are invitations to sit on journal editorial boards, speak at prestigious conferences, and ultimately achieve the coveted status of tenure. This career path produces continuity in science, but it inhibits innovation and the creative spirit. Bright young minds are reigned in to conform to the established areas of "credible research" choices. The prospect of leveraging off your postdoctoral mentor is indeed difficult to resist but is often the first step to maverick status when one does. Venturing outside those boundaries can, however, slow a scientist's progress up the career ladder since the mentors advocacy tends to diminish when the student ventures into new territory. The intellectual correctness of such career paths relegates ideas to a caste system in which innovation becomes not only less important, but also potentially dangerous. This is a key component of the politics of science, and grant review boards tend to marginalize maverick investigators when they venture outside the intellectually correct box that thought leaders often create.

An alternate and much less followed path is for the scientist to step out of this conformity and challenge conventional wisdom. This approach is much riskier and offers its practitioner much less funding security since it lacks the leverage of the thought leader and their body of work. The credibility of the mentor's work is established, whereas that of the budding scientist is not, so they are on their own. Getting the first grant is tougher and this always plays a pivotal role in getting tenure. This rare breed is the maverick scientist.

Mavericks spend a lot of time swimming upstream. They exist on the edges of the massive science establishment; they have no constituency within the science community and no platform for their ideas. They are not interested in the conventional wisdom and are generally open to using knowledge already available to achieve their goal. They come up with the "outside the box" answers to very difficult problems that have persistently eluded traditional approaches. Curing the terminal diseases is certainly one of those situations. The thought leaders are not doing such a good job, and it would appear that after 50 years of trying to crack open the mysteries of terminal disease that the maverick mentality may be warranted.

Maverick science can take many forms. The ability to look at old things in new ways transcends both basic and applied research and is what innovation is all about. The dogma regarding cancer, degenerative disease, and cardiovascular disease has failed to provide cures for 50 years, which should underscore the need for mavericks. It is amazing when one compares the success of maverick science to mainstream science over the past several decades that the mavericks are doing much better in spite of the huge support mainstream science has garnered. There are several cures that have emerged from the innovation of mavericks, while organized science has little to offer by way of comparison.

A great example of maverick science is the pioneering accomplishment of Alexander Fleming in 1928, regarding the most impressive medical advance of the past century. He observed that some of his Petri dishes containing bacteria had areas in them where the bacteria were not growing. He wondered whether something might be inhibiting their growth. Then he observed that some mold was growing on the plates. This was a chance observation, but it was made by an unbiased mind that was sufficiently open to ask the next question: Might the mold be killing the bacteria and if so, how? The traditional scientist of the day might have completely ignored what Fleming observed, since it made no sense in the context of then contemporary science. Mold is everywhere as are bacteria so why should this happen? Fleming postulated that the mold was making something that was selectively killing the bacteria and decided to figure out what that might be. It took substantial humility and an element of romanticism to take this course. The more arrogant scientist tends to write off what does not make sense and not chase dreams. If it does not fit a theory, forget it because it does not exist.

It took years for Fleming to convince his superiors that he was not crazy and to mobilize the chemists necessary (he was a bacteriologist, not a chemist) to actually isolate and characterize the active principle that he hypothesized the mold was making. This would have happened much sooner if a thought leader of the day made the same observation, but thanks to Fleming's persistence, it was eventually determined that the active principle was a substance we now know as penicillin, the single most important medical advance of the 20th century. This advance cured a host of what were then fatal infectious diseases. It occurred without public funding, elaborate infrastructure, or the support of organized science. One stubborn maverick, innovation, persistence, and hard work were the only requirements.

There are not many scientists today who would stoop to the pure and simple observation of a mold, bacterial, herbal, or other types of living or nonliving soup, to find substances that would selectively kill cancer cells or inhibit nerve cell degeneration or destroy the AIDS virus. Today's academic is above that sort of opportunistic and pragmatic simple science. They have evolved to technocrats and in the process lost the courage to innovate and search for new leads where the light has yet to shine.

Fleming's discovery was a revolution in medicine since infectious diseases were, in many cases, fatal before the discovery of antibiotics. Infectious disease was a major cause of death back then, while today it is routinely cured with a simple and inexpensive prescription. Fleming's discovery did not come from a research project that played out over decades and cost millions of dollars, yet it represented a cure for an entire class of disease.

A more recent example of maverick science that occurred in the 1980s is the revolutionary discovery of Drs. Barry Marshall and Robin Warren in Perth, Australia. Again as with Fleming, neither was a thought leader or had huge research grant funding and they framed a basic hypothesis that was outside the boundaries of contemporary research dogma. They then proceeded to test and prove it without any substantial support from organized science. It led, almost immediately, and much to the embarrassment of the entire academic-pharmaceutical industrial complex, to a cure for peptic ulcers [1,2].

For decades, the prevailing belief was that peptic ulcers were caused by excess stomach acid production triggered by anxiety and that this excess acid destroyed parts of the stomach lining and produced life-threatening bleeding ulcers. Therefore, blocking this acid secretion (a perfectly natural process) with acid blocking drugs emerged as the therapy of choice. A multibillion dollar industry developed around several acid blocking drugs (Tagamet and Zantac), and the medical establishment was quite proud of itself even though it was apparent that this approach did not cure ulcers but only treated the symptoms.

But there was one small problem that the thought leaders in both academia and the pharmaceutical industry had not thought of. The acid-blocking drugs were based on a false assumption of what actually *caused* ulcers. It was not excess acid; that was only irritating the existing ulcer. The cause of the ulcer was something entirely different. They were actually only treating the symptoms. These two drugs were, however, very profitable and they did effectively treat the symptoms of ulcers. They also had to be taken for long periods of time on a daily basis, a characteristic that the drug industry gets very excited about for obvious reasons. The science community was also of one voice on the cause and treatment of peptic ulcers. For all intents and purposes the medical and pharmaceutical community thought they had this disease pretty much under control and in a manner that provided a very steady and attractive cash flow. That was all rapidly turned upside down by the two maverick investigators.

Drs. Marshall and Warren were marching to a different drummer, one that is typical for mavericks and at odds with thought leader dictated medical science. They stunned the medical community with the proposal that peptic ulcers were actually caused by a bacterial infection of the stomach wall and not by excess acid secretion. No one in the science community took them seriously since it was thought that *nothing* could live in the strongly acidic environment of the stomach. To prove their conviction, one of the physicians actually ingested some of the bacteria (*Helicobacter pylori*) that they suspected caused the ulcerous stomach lesion [1]. He, in fact, developed ulcers and then proceeded to cure his own case with a simple antibiotic. The fact that such a dramatic gesture was required illustrates the power of thought leader induced intellectual correctness in the politics of science. How many other "well-accepted" pieces of science dogma are there which may be flat out wrong? Might other cures be lurking in unsuspected places? We will never know were it not for mavericks.

More rigorous studies followed and proved in a very short period of time that existing off-the-shelf antibiotics, such as erythromycin or amoxicillin to name a few, could not only treat peptic ulcers far better than the acid blockers, but that these old drugs actually cured ulcers. The stomach wall proceeded to heal and precluded the need for ulcer surgery. The entire process to prove this cure required very little investment, no infrastructure-laden companies, universities, or research institutes. It also did not involve developing new drugs but rather using old ones in new ways. Innovation and the courage to venture outside the boundaries of intellectual correctness were all that was necessary. This was not only an embarrassment to the entire medical R&D establishment, but also a profound illustration of how formidable its resistance to outside innovation can be. Obviously not much money was made on this cure since a new drug or patented product was not involved. This is probably why it escaped the drug industry for such a long time. They do not spend time on projects that have no earnings potential.

It is interesting that so little attention was given to this event. The media reported it, but there was no analysis of what it was trying to tell us about the cumbersome regimented closed-minded politics of the R&D process. Industry was presumably embarrassed by the discovery and academia pretty much did not care since the cure involved old science that they could not take the credit for. This incident should have started a national dialog about the medical discovery process. Questions such as how it might be possible to facilitate the use of old, unpatented drugs for new uses should have been addressed. They were not since it was something both the industry and the academic community either wanted to forget or did not, for different reasons, care about in the first place.

It is interesting to consider the contrast between the cumulative successes of organized science in finding cures with that of maverick basic scientists. The former has had enormous support from foundations, granting sources, and academia, while the later operates in the shadows. The mavericks are quick and generally look at problems directly and not through the lens of science dogma. They are oblivious to their peers and in fact must fight such influence to innovate. Science is a mix of iterative process and insight. To the extent that the iterative prevails, the cures will be slow in coming.

OLD DRUGS FOR NEW USES AND MAVERICKS

There are thousands of drugs that have been developed. Knowing what a drug does in one disease is often the tip of the iceberg. Also, old drugs have a known safety profile. Drug companies often do not want to venture into other diseases with their drugs for fear that they may find out something that might jeopardize the drug in its current use. As we shall see later, it cost Merck over 7 billion dollars in legal claims and an eventual class action law suit settlement when it tried to demonstrate the benefit of the pain drug Vioxx in colon cancer patients. That 18-month study showed cardiovascular side effects in these patients when Vioxx was administered for that extended period. In spite of these unfortunate stories, and there are many, as we will see in Chapter 6, mavericks still exist. Imagine if they were supported and encouraged by the R&D establishment rather than marginalized, it would open up huge untapped talent and give innovation a big boost.

Thalidomide is probably the most infamous drug known. Originally developed in Germany in the late 1950s, it was never formerly approved in the United States, but was marketed throughout Europe and eventually found its way to a limited number of patients in the United States. The drug was very effective as a sedative for pregnant women. Sadly, it was quickly discovered to have profound side effects in the babies born to mothers using the drug. Terrible limb deformities were seen in many of the babies, even if the mother took only one dose of the drug at the appropriate time during pregnancy.

Thalidomide-induced birth defects caused a firestorm throughout Europe and the United States in the early 1960s, and the drug was promptly banned. Although it was never officially approved in the United States, thanks to one very bright and dedicated drug reviewer at the Food and Drug Administration (FDA), Francis Oldham Kelsey [3], but it was given to some patients in the United States. In spite of the FDA acting appropriately by not approving the drug, the thalidomide incident induced Congress to broaden the authority of the FDA and mandate that drug companies had to prove that drugs were not only safe but also effective. After such a calamity, the last thing one might expect was to see thalidomide become a "wonder drug." But thanks to the efforts of some maverick investigators, that is precisely what happened.

An Israeli physician, Jacob Sheskin, was looking for a way to sedate a very agitated leprosy patient. He gave the patient some thalidomide since he had nothing else at his disposal and the patient was not a pregnant female. The drug sedated the patient as expected but he also noticed that the skin rash associated with leprosy was greatly diminished by the drug. Again as with the Fleming example, a more traditional scientist would have dismissed the observation as a chance occurrence since there was no rational explanation for it. Dr. Sheskin, however, in true maverick fashion, continued to investigate whether thalidomide could have new uses other than as a sedative. There were not many scientists willing to even touch this drug at the time even though the only caveat associated with its use was keeping it away from pregnant women. Eventually, his observations resulted in a clinical trial of the drug in leprosy patients where it was shown to be a potent antiinflammatory agent that cured the skin lesions characteristic of the disease.

Years later, a small start-up biotech company, whose founding technology was failing, decided to reorganize itself and take a flyer on thalidomide. Companies can be mavericks too. The company was called Celgene, and at the time it had few alternatives other than an old drug–new use strategy. Celgene did the necessary clinical testing and received formal FDA approval for the treatment of leprosy in 1998. In the years since, virtually all the leprosy hospitals in the world have been closed thanks to the curative effect of thalidomide on this autoimmune inflammatory disease. Thalidomide, a once feared drug, became a cure for a disease because a maverick physician had the insight to transcend the dictates of traditional science. Under today's healthcare rules this would never be permitted and effectively eliminates a valuable asset from the cure effort, the maverick physician.

The leprosy approval stimulated further interest in thalidomide for other autoimmune diseases and prompted studies on the mechanism by which the drug worked. Thalidomide has since been shown to be virtually a wonder drug [4] with multiple therapeutic actions, including antiinflammatory efficacy and inhibition of blood vessel growth, a process called angiogenesis that facilitates cancer tumor growth. Consequently, when combined with chemotherapy drugs, thalidomide has been shown to be an effective anticancer agent. In 2003, thalidomide was approved in Australia and New Zealand for the treatment of multiple myeloma. It is likely that thalidomide and its derivatives will have a major impact on many serious diseases in addition to leprosy. Remember, this is not only old science but also condemned science that when looked at with a new and innovative eye in the hands of a maverick, rapidly advanced medicine.

You may be asking why the lack of patents on thalidomide did not impede its commercialization. In fact it did, but the effect was not as great as it might have been if the drug was still marketed for its original use. Remember, the drug was pulled off the market back in 1962. Also, the developers of the drug have made novel derivatives that are patented in an attempt to improve the clinical characteristics of the drug for the new indications.

Yet another example of how old drugs can be used in new ways is aspirin. It has been in use as a pain reliever for over 80 years and for the last 30 years has been known to be an effective inhibitor of blood clotting. This came largely from the observation that aspirin increased bleeding. Physicians began to notice that patients who took aspirin regularly had reduced blood clotting and less acute cardiovascular events, such as heart attacks and strokes. This led to clinical trials of aspirin therapy. Aspirin is now used on a prophylactic basis to prevent heart attacks induced by blood clotting. It is also used acutely to stop heart attacks after they occur. A low dose of aspirin daily can reduce the incidence of heart attacks and strokes by up to 30%. Aspirin therapy probably saves and improves more lives than many of the much more expensive drugs such as the enormous variety of blood pressure and clot busting drugs that now clutter the market. The added benefit of this and all other old drug–new use scenarios is that they are quick, safe, and very inexpensive to develop.

GETTING PHYSICIANS BACK IN THE GAME

The aspirin example as well as the peptic ulcer and thalidomide examples underscores the power of physician observation in the generation of new therapies. If physicians take the time to follow their patients and see how they respond to drugs, they can come up with important insights regarding other beneficial effects of a drug. Often, physicians have hunches about drugs and will use them in an "off-label" manner. This means using a drug for a disease other than that for which it was intended and approved for. Off-label use of drugs is not generally reimbursed by healthcare, and this is unfortunate since it inhibits an important outlet for physician innovation. It also exposes them to litigation, a reality that has forced them to employ defensive medicine rather than innovate. This is especially unfortunate when terminal patients are involved.

Physicians are the real-world practitioners of medicine and, as we have seen, can be an important source of maverick science since they understand drugs and illnesses and are able to generate and test new treatment insights in ways that basic scientists and drug companies cannot. It is unfortunate that the physician–patient relationship has been so compromised in contemporary healthcare. Physicians have become employees of Health Maintenance Organizations (HMOs) or insurance companies at an increasing rate that will soon take them totally out of the arena of drug innovation. This important research opportunity for testing approved drugs is being wasted when it should be encouraged. There are thousands of safe and effective approved drugs to choose from, and when one considers combinations of such drugs, the potential for innovation becomes endless. This source of raw material has not been mined, and physicians are the appropriate miners but too shacked by the healthcare system to pursue this opportunity. Reform of the healthcare system is essential to remedy this problem and will be discussed in Chapter 11.

NATURAL AGENTS ARE BEING OVERLOOKED

Maverick science has also been applied to natural substances to treat disease; unfortunately not nearly enough of this has occurred and the industry marginalizes it for obvious reasons; patents on natural substances are difficult to obtain. The major obstacle here is again the thought leaders and an academic mentality regarding the preference for novel cutting edge science rather than natural product medicine. The particular example I will use relates to neuroprotective drugs, an area that I am very familiar with. These are drugs that could make a huge difference for stroke and head injury patients for which there are currently no treatments. The strategy now is simply to wait and let the patient recover naturally so it may be that supplementing with natural substances would improve recovery. Old drugs used in new ways do not get some attention of mainstream science, but naturally occurring substances suffer even more in this respect. Academics do not get excited about this because it is not new science, and the industry is also not interested simply because they cannot make money using natural, and hence, unpatentable molecules. This, however, makes them more intriguing to the maverick that prefers to look where no one else is.

Neuroprotection has long been an elusive goal of neuroscience. Drugs that could protect the human brain during and immediately after periods of acute injury from a stroke or head trauma would not only save many lives but also improve the quality of life for millions of people. Research in this area over the past 40 years has been extensive and much is now known about the mechanism of nerve cell death, but an effective treatment for this major unmet medical need is still nowhere on the horizon. Not only would neuroprotective drugs be important for acute brain injury but they would also possibly serve to stop the degenerative progression in Alzheimer's and Parkinson's disease as well as a host of other neurodegenerative diseases. The pathology is probably very similar for both acute and chronic neurodegeneration and may differ only in the speed by which they occur. The failure to produce neuroprotective agents has been a key part of the cure crisis. There has been a lot of science but no progress toward cures.

Back in the 1980s, a young maverick scientist, Donald G. Stein, made an observation that his peers chose to ignore. He observed (by chance) that some female animals whose brain was injured seemed to recover better than males. None of the males recovered but most of the females did. The observation intrigued him but no one thought something associated with being female could play a role in mediating CNS repair, especially since all the females were not protected. Dr. Stein, however, being a maverick, was not wedded to intellectual correctness and persisted in his study of potential sex differences in the recovery from traumatic brain injury. He found that females were more resistant to brain damage in certain periods of their reproductive cycle. It took Dr. Stein decades to get other researchers in the field to take him seriously. Finally, after many years, much perseverance, and eventual corroboration by other investigators, he was taken seriously. So, what was it that he discovered?

Dr. Stein looked at the simplest things first: the female hormones. He tried giving estrogen to brain-injured male rats with no beneficial effect. He then tried progesterone and that was very effective. Years went by with both the basic science and industry folks remaining unimpressed. The technology was not new and attractive as some of the other targets being pursued at the time, and there was no known mechanistic rationale for the effect. The patents would be weak since progesterone is a natural substance. To make it even worst, progesterone was already available commercially to manage birth control and other female hormonal problems. This was an old drug–new use scenario with a twist.

Dr. Stein's discovery was going nowhere because it did not fit into the typical drug development paradigm. Fast forward to today. Based on almost 400 preclinical studies from around the world, the progesterone theory of neuroprotection has gained acceptance and led to several academic human clinical trials in head injury patients, some of which showed benefits and others which did not due to dosing irregularities and study design issues [5–8]. Academic clinical trials often are wrought with such irregularities because they are generally not executed under the strict quality control guidelines employed in the drug industry. This is a major problem and includes trials conducted by the NIH where one of the progesterone trials was performed. This issue will be discussed in later chapters where reforms to the NIH are proposed. It will require a drug company functioning under strict industry standards to demonstrate the efficacy required to justify approval of progesterone for neuroprotective indications. This is not likely to happen due to the fact that enforceable patents on progesterone therapy are not possible to obtain; an issue we will also discuss in later chapters on patent reform.

While it is unclear that progesterone will be a viable treatment for brain injury and stroke patients, it is clear that were it not for a determined and innovative maverick scientist with the courage, or perhaps the stubbornness, to swim upstream, this approach to neuroprotection would never have materialized. Mobilizing and creating incentives for old drug–new use and natural substance evaluation would open up a very important innovation arena. If this approach was facilitated rather than shunned, it could become an important part of the quest for cures rather than the curiosity it currently is. Mavericks must be encouraged, especially when it comes to curing diseases that have eluded all efforts for such a long time.

CHARITIES: AN INSULT TO THE CURE QUEST

Let us go on a walk to cure cancer or contribute to a charity to cure Alzheimer's. Have these walks done any good? Are we so desperate to trust that this will actually make a difference? Name one that has come up with anything that even comes close to developing any treatment for anything. The reality is that engaging in these futile efforts that share the same problems discussed in the previous chapters is an expression of our empathy to the victims of terminal disease but that is about all. Charities mean well and some do provide support for those with terminal disease, but their record as far as curing such diseases is not good.

THE BOTTOM LINE ON THE RESEARCH ENTERPRISE

The past two chapters and this chapter paint a picture of an enormous medical research infrastructure characterized by an absolute absence of coordination, strategy, and accountability where science is the focus and cures a fringe benefit. The academic community is assuming a role it was not designed to perform and too few thought leaders are leading the charge to cure disease. Accountability for accomplishing the mission is totally absent. The NIH lacks the leadership required to accomplish the task of even doing a viable clinical trial, let alone managing an effort as urgent as curing terminal disease. Universities and research institutes employ no quality control and no metrics for either fiscal or results accountability. Researchers are motivated by metrics (tenure, academic freedom, and publication) that have nothing to do with curing disease.

It has now, after some 60 years of effort and trillions spent, become crystal clear that we are witness to an enormous human tragedy. Change must occur if this fiasco is to be corrected and the first step is recognizing the problems. Hopefully the past chapters have shed some light on the fundamental deficiencies that characterize the research side of the problem and the futility of simply pumping more money into it in hopes that something relevant to cures will fall out of it. The good news is that the problems are fixable and do not require increased cost but rather a more appropriate deployment of existing infrastructure in a manner appropriate to the task. Before going there the development, regulatory, and legal aspects of the cure enterprise must be exposed since they are in as bad shape as the research arena.

REFERENCES

[1] J. Hardy, Medical wisdom challenged by a cocktail, Hardy Diagnostics. Retrieved from: http://www.hardydiagnostics.com/wp-content/uploads/2016/05/Barry-Marshall-H.pylori2.pdf.

[2] B. Marshall, P.C. Adams, *Helicobacter pylori*: a Nobel pursuit? Canadian Journal of Gastroenterology and Hepatology 22 (11) (November 22, 2008) 895–896. Retrieved from: https://www.ncbi.nlm.nih.gov/pmc/articles/PMC2661189/.

[3] R.D. Mcfadden, Francis Oldham Kelsey, who saved U.S. babies from thalidomide, dies at 101, New York Times (August 7, 2015). Retrieved from: http://www.nytimes.com/2015/08/08/science/frances-oldham-kelsey-fda-doctor-who-exposed-danger-of-thalidomide-dies-at-101.html?_r=0.

[4] G. Greenstone, Special feature: the revival of thalidomide: from tragedy to therapy, BC Medical Journal 53 (5) (June 2011) 230–233. Retrieved from: http://www.bcmj.org/newsnotes/special-feature-revival-thalidomide-tragedy-therapy.

[5] D.W. Wright, et al., ProTECT: a randomized clinical trial of progesterone for traumatic brain injury, Annals of Emergency Medicine 49 (2007) 391–402, http://dx.doi.org/10.1016/j.annemergmed.2006.07.932.

[6] R.B. Howard, I. Sayeed, D. Stein, Suboptimal dosing parameters as possible factors in the negative Phase III clinical trials of progesterone in TBI, Journal of Neurotrauma (2015) http://dx.doi.org/10.1089/neu.2015.4179. (E pub ahead of print).

[7] D.W. Wright, S.D. Yeatts, R. Silbergleit, et al., Very early administration of progesterone for acute traumatic brain injury, The New England Journal of Medicine 371 (26) (2014) 2457–2466.

[8] B.E. Skolnick, A.I. Maas, R.K. Narayan, et al., A clinical trial of progesterone for severe traumatic brain injury, The New England Journal of Medicine 371 (26) (2014) 2467–2476.

Chapter 4

The Strangled Unincentivized Drug Industry

Chapter Outline

As with academia, the pharmaceutical industry is failing at cures. The reasons are many but primary among them are their lack of incentives to focus on cures, focus on me-too drugs, a hostile Food and Drug Administration (FDA), and an out-of-control medical legal system. The biotech industry shares those problems and also suffers, to some extent, from the same infatuation with science rather than cures that academia does. Both the pharmaceutical and biotech sectors also suffer from the thought leader syndrome that is so pervasive in academia and has greatly inhibited innovation. Before discussing these problems it is essential to understand the difference between pharmaceuticals and biotechnology.

BIOTECHNOLOGY VERSUS PHARMACEUTICALS, MACRO VERSUS MICRO

The drug industry has changed dramatically since 1970. Up until then it consisted primarily of the large multinational pharmaceutical companies that were developing small molecule drugs (pharmaceuticals) based on leads from natural plant sources as well as organic chemistry and biochemistry research. The entire industry was pharmaceutical in nature, which means it focused on small molecule drugs that could be made by chemists in the laboratory. Natural product macromolecules such as proteins and hormones were used but they had to be

A Roadmap for Curing Cancer, Alzheimer's, and Cardiovascular Disease.
http://dx.doi.org/10.1016/B978-0-12-812796-4.00004-0

purified from harvested tissue, which was extremely laborious and difficult to manage regarding quality control. That changed in the 1970s with the advent of biotechnology, which led to hundreds of small research-based venture capital backed start-up biotech companies such as Amgen, Genentech, and Hybritech.

The next two decades saw an enormous proliferation of these entrepreneurial biotechnology companies. Biotech companies generally get their start from academic founders and the financial support of venture capital firms that typically end up running and owning large portions of the company. This explosion of small academically oriented biotech companies has had profound effects on the drug development process. Unfortunately their effect on the delivery of healthcare and the curing of disease has been minimal, and the cost of the few drugs that have emerged has been extreme.

There is a fundamental difference between pharmaceutical drugs and biotechnology drugs or biologics. The two strategies of drug development are quite different with regard to the basic aspects of drug delivery to the disease site and a host of other important parameters for drug action. Pharmaceutical companies use small synthetic molecules to target various naturally occurring proteins in the body, while biotechnology companies use these proteins themselves as drugs. Biotechnology is really quite simple in concept, yet few lay people or public policy makers really understand the ramifications of the raw material of biotechnology (proteins) and how difficult they are to produce, formulate, and deliver to the appropriate place in the body. If this was made clear 40 years ago, biotechnology would have received far less emphasis in drug development. No one was really thinking strategically back then, or even now, and we let science drive the boat rather than cures. Had there been a plan, biotechnology would probably have taken a different place in the industry than is currently the case.

Biologicals are almost always very large molecules, such as enzymes or antibodies that are composed mostly of protein. Often these enzymes and antibodies are associated with sugar molecules (glycoproteins) or with fats (lipoproteins). These proteins, glycoproteins, and lipoproteins are all enormous in size and complexity when compared with pharmaceuticals. For this reason they are impossible to make in the traditional chemistry laboratory and are therefore very difficult to use as drugs since they have to be purified from living tissue, a very expensive and imprecise task.

The innovation that biotechnology provided was a means of actually making proteins in the laboratory outside the human body. This is done by taking the genes that make specific proteins from the cells in question and putting them into bacterial cells, which can be grown in the lab by fermentation. The protein of interest is then isolated in pure form. Biotechnology is simply a way to make something in the laboratory, which we already knew existed, so it is a technique and not new knowledge about how cells work and what goes wrong in a disease. Therefore, if a disease is caused by a defect in a specific protein, one can now make the correct protein, give it to the patient, and theoretically manage the disease. At first blush this seems to be an important advance for the drug industry, but unfortunately the wellspring of cures that was touted back in the 1970s has not materialized. The reasons for this are manifold, and all of them were common knowledge back in the 1970s and should have been realized. Unfortunately science trumped strategy, and yet again this lack of a strategy caused precious time and resources to be wasted. So what are the problems with biotechnology?

BIOLOGICS: GREAT TARGETS BUT TERRIBLE DRUGS

Issue **one** with biotechnology is the enormous size of proteins; we call them macromolecules. This presents major difficulties in the dynamics of drug delivery to the disease site. A drug must get to the disease site unaltered if it is to have an effect. The industry jargon for that critical site is the "therapeutic target." The body is made up of many compartments and neighborhoods in which therapeutic targets reside, and transit in and out of these places is highly regulated. The size of proteins quite simply precludes their entry into many of the body's compartments, and this limits their utility dramatically. A good analogy might be attempting to navigate a narrow mountain footpath in a school bus or a mountain stream with an aircraft carrier. The logistics of getting to the therapeutic target with proteins is a major issue that was not spoken about a lot when the biotech companies proliferated. Size and access to the therapeutic target is therefore the first major limitation of biotechnology drugs.

Proteins are extremely complex molecules that are composed of superlong strings of amino acids that are hooked together in a very specific sequence. This sequence is dictated by the DNA in the gene that codes for and determines which of the 20 amino acids go where in a given protein. The sequence of amino acids in a protein partially determines what it does in the body, and often times it is an aberrant sequence of amino acids in a given protein that causes a disease. The protein's aberrant structure is therefore the therapeutic target. Unfortunately, biotechnology only enables the stringing of the amino acids together in the proper sequence. I say unfortunate because there are many proteins that require more than just their amino acid sequence to impart their full function. Many human proteins and enzymes require some sugar molecules (glycoproteins) or fatty acid molecules (lipoproteins) attached to them, as we discussed previously, to be biologically active. Some proteins require both. When it comes to these proteins, biotechnology is compromised in recreating a totally faithful replica of the natural enzyme or antibody. This is the **second** major limitation of proteins as drugs or biologics; many proteins are not good candidates for its use.

The **third** limitation of biotechnology is that many enzymes contain two, four, or more chains of amino acids or subunits, all hooked together in an elaborate superstructure. Biotechnology is only capable of making one chain and thus can only be optimally utilized with proteins that exist in nature as a single amino acid chain. With biologicals one has to be quite selective with not only the location of the therapeutic target but also with the protein candidate that will serve as the drug. This narrows the playing field for biotechnology dramatically. But there are further limitations.

Enzymes, antibodies, and proteins are among the largest and most structurally elegant molecules in the universe. They are orders of magnitude larger than pharmaceuticals and have to be in a very specific three-dimensional conformation to serve their biological function. The string of amino acids folds in a very specific way and must keep that folded shape to be active. A useful way to visualize this and get a sense of the complexities of trying to make biologicals into drugs is to think of a drug as a length of string. Let us say for the sake of comparison that a pharmaceutical drug would be about an inch-long piece, while a biological would range anywhere from 10 to a 100 feet long! Add to that the requirement is that the string must have certain two- and three-dimensional shapes or folding conformations with respect to itself to be biologically active, and you have a very complex and delicate drug. The pharmaceutical at one inch long is relatively easy to keep in the active conformation, but the 100 foot long string is a nightmare to maintain in its active conformation during delivery to the therapeutic target. How does one get that enormously fragile entity to the therapeutic site in the active conformation? Certainly not in a pill that can be taken every day. Stability is therefore the **fourth** major problem with biologicals.

It is virtually impossible to deliver a biological as a pill. The delicate, highly folded elaborate structure would be destroyed in making the pill and rendered useless. For this reason, almost all biologics are delivered intravenously where they can gain immediate access to the bloodstream. Intravenous delivery is not a major problem, but unfortunately there is yet another problem (limitation number **five**) to consider with biotechnology drugs. Their extreme size and complex conformation make them very difficult and, in almost all cases, impossible to get the biologic out of the bloodstream and into cells. Tiny pharmaceutical drugs generally have a much easier time exiting the blood and getting inside cells; they have a much better bioavailability profile than biologics. If the therapeutic target is not in the bloodstream, biologic drugs are of very limited, if any value. This generally means that either the therapeutic target must be something in the blood, such as a blood clot or a blood cell, or something on the outside surface of a cell that is directly accessible to the blood. This further limits biologicals since many therapeutic targets are not directly accessible to the bloodstream.

There is yet another limitation of biologicals (number **six**) as therapeutic entities. Certain organs in the body restrict what they will permit to enter their cells. Primary among these is the brain. Brain blood vessels are insulated with an almost impenetrable layer of tissue that prohibits almost everything above a certain molecular size from entering the brain. This well-studied structure is called the "blood–brain barrier" (BBB). Obviously the BBB in almost all cases excludes proteins from gaining access to the brain. Pharmaceuticals share this problem but to a much lesser degree. It is unlikely that any drug that needs to gain access to the brain will ever be a protein unless a specific transport system exists or the BBB is temporarily compromised. This, at least theoretically, means that biologics are unlikely to be cures for the myriad of terminal brain diseases, such as Alzheimer's, Parkinson's, amyotrophic lateral sclerosis, brain tumors, and many others, unless they are one-time treatments that can be surgically implanted or the BBB is compromised.

Biologics are large molecules and the immune system generally targets large molecules. Repeated exposure to biologics can therefore trigger an immune response, which destroys the biologic and results in the loss of efficacy, stopping the biologic and switching to another one if it is available. This can to some extent be managed by coadministration of immune-suppressant drugs but that does, however, sometimes cause additional problems depending on the biologic. So here we have the **seventh** limitation for biologics.

Biologics are very hard to copy. It is very difficult to make generic versions of such large molecules, and the FDA has been having real problems coming up with metrics to evaluate these "biosimilars" as generic biologicals are called. The industry is of course thrilled about this since it is good for their bottom line because the presence of generics forces them to cut their price by 50% or more. This is problem number **eight** for biologics.

Finally, biologics are extremely expensive. Making and vetting biologics is a costly process. They must be made in sterile clean rooms since any bacterial contamination could cause sepsis when given intravenously, which is the usual route of administration for biologics. Most have to be stored under refrigeration or as freeze-dried vials. All this adds enormous cost and results in finisher products that can cost tens of thousands of dollars per dose. So, problem number **nine** is that biologics are wreaking havoc with healthcare insurance premiums and Medicare/Medicaid costs.

BIOLOGICALS: SUPER EXPENSE FOR MARGINAL BENEFIT

Biologics are clearly plagued with many limitations when used as drugs and one wonders why anyone would bother with them. Certainly they can be useful for some therapeutic targets, but it is difficult to imagine that they will cure cancer or neurological diseases. Pharmaceuticals seem to be an easier way to approach disease. Biotechnologists have been forced to be very selective in choosing applications of their drugs. They have to find a disease for their technology rather than a technology for the disease. It typifies a key problem in the quest for cures, an inordinate focus on technology/science rather than the disease itself, and a lack of strategic planning.

Some examples of how biotechnology has tiptoed around the enormous limitations of their technology to muscle out drugs are cardiovascular disease for dissolving blood clots (tissue plasminogen activator), diabetes (human insulin), anemia (Epogen to stimulate red blood cell growth), and various autoimmune diseases such as Enbrel and Remicade. There are also some biologicals for cancer, such as Rituxin, but they also target the immune system or blood vessel growth (antiangiogenesis) and basically augment the body's ability to fight cancer when used with the older chemotherapy drugs, which are pharmaceuticals. All these biotech drugs act on therapeutic targets that are to varying degrees accessible to the bloodstream, and all are intravenous infusions. None cure the disease they treat, and their prohibitive cost has threatened to bankrupt the entire healthcare system with treatment courses in the six figure range. Why has this been allowed to happen? Well one reason is that the cost of biologics is enormous, and making generics is not easy given the complexity of the molecules involved. This certainly makes biotech companies smile but has not been kind to the healthcare budgets of families. The most simple-minded of strategies would have argued against the widespread use of biotechnology *40 years ago* since all of the limitations cited for biotech drugs were common knowledge back then. A strategic approach to medical R&D would have prevented that.

GENE THERAPY AND STEM CELL THERAPY: HYPE VERSUS SUBSTANCE?

Perhaps two of the most aggressive potential applications of biotechnology are in the areas of gene therapy and stem cell therapy. We have seen how difficult it is to make fully active proteins and deliver them to a therapeutic target, but in spite of this the idea emerged to try and insert the gene that makes the protein (gene therapy), or replace the entire defective cell (stem cell therapy). In many ways this is like going from the frying pan into the fire. Proteins are big and complex, but a gene is much bigger, and a cell is orders of magnitude larger and complex than both of them combined. Given the problems that were encountered with proteins, it seems bizarre that an even more difficult feat would be attempted, but there seems to be no limit to the faith that virgin science instills especially when the media does not seem to understand these basic points and gives it front page status as potential cures for everything.

To deliver genes into the nucleus of human cells, one needs to employ a means of accessing the nucleus and then actually inserting the gene into the enormously complex chromosome structure we still do not fully understand. To do this researchers have utilized viruses, which naturally have the ability to insert their genetic material into human chromosomes. The human gene in question is inserted into the viral genome using biotechnology techniques, and the virus acts as a delivery entity, or vector, to deliver the gene into the patient. While the science involved in gene therapy is extraordinary, it does not take an extensive knowledge of biology to fathom the difficulties of such an approach. As with biotechnology, the practicality of gene therapy is not encouraging if one considers what actually has to be accomplished. Not only does the gene have to be *delivered*, but the defective gene should also be *removed* or skipped as well, something that is very complex. The regulatory mechanism for expressing the gene must also be maintained, a process we also do not fully understand. Also, the gene must get to the right cells, the viral vector must be controlled, and the consequences of delivering too many or too few copies of the gene dealt with. Also, if a drug does no work or causes side effects one can simply stop taking it. With gene therapy there is no reversing it, something that should preclude the entire strategy. The technical hurdles with gene therapy are therefore monumental and the potential for long term side effects unknown. The assumptions required to predict success with gene therapy are mostly untested and unknown, faith plays a role and it should not.

Stem cell therapy; the most recent biotechnology approach to disease therapeutics has captured the imagination of virtually every aspect of the medical establishment, and the public, thanks to effusive hyping by the media, celebrities, and politicians that have jumped on the band wagon. What stem cell therapy promises is that if a certain type of cell in a specific organ is destroyed (as occurs in degenerative diseases), it can be *replaced* by providing that organ with a stem cell population that can become that cell. Stem cells are precursors to differentiated cells. They have the ability to develop or differentiate themselves into several different cell types, such as nerve, liver, kidney cells, etc.

The simplistic notion behind this extremely complex strategy is that the stem cell will develop into a nerve cell in the brain of a Parkinson's or Alzheimer's patient and miraculously correct the impairment associated with the disease. This combination of new, and again, virgin science coupled with past failures regarding cures has led to a "frenzy of hope" in stem cells. What is not often mentioned is what we know about basic developmental biology, which precludes the premature application of this technology. For example, it requires a leap of faith to assume that randomly generated nerve cells in an adult brain will miraculously assemble themselves into the complex structure required to restore normal brain function. How will the stem cell differentiation be regulated? Will these new cells overgrow existing cells in other nondiseased areas of the brain? Will these cells form tumors as has been suggested [1,2] since stem cells and cancer cells share many biological characteristics?

There is another very troubling aspect of stem cell therapy. As with gene therapy it is essentially irreversible. The stem cell becomes part of the body and a potential time bomb. Rogue cells with unknown effects might arise that could create problems more serious than the disease originally targeted. The delivery issues are also difficult as are the immunological problems. In many ways employing stem cell therapy is like playing with fire and will create safety issues that will challenge the FDA and be a picnic for the legal community as they exploit the liability issues.

Stem cells have been hyped by thought leaders much like biotechnology was several decades ago and gene therapy after that. It has been an endless cycle of excitement about advances in science but, after several decades, virtually no treatment breakthroughs in the diseases that we are helpless against. If the thought leaders can fan the flames of hope and anticipation, it butters their bread in the form of attracting funding to their work and building their empires with the help of venture capitalists building companies based on pure hype. No one will remember when the failures start to appear years later, and there will be yet another body of preliminary unvetted science to step in and take its place.

Stem cells and gene therapy should be explored as legitimate science for their own sake, but the frenzy that has developed around them as potential cures is entirely unwarranted and serves to siphon talent and resources from other more near-term approaches. The best evidence for this is past history. California appropriated three billion dollars for stem cell research in 2004 before any rational use of the funds was even proposed [3]. Others are following that example with no expectation of what the funds can produce or when. More importantly, the money is being pumped into the same unaccountable disorganized system with all the problems previously discussed.

Chasing virgin science is a windfall for academia since they are the beneficiaries of the spastic infusions of capital. The enormous capital infusions into these hyped strategies would be better spent on turning over more stones with already vetted science.

THE BIOTECHNOLOGY REPORT CARD

The distinction between pharmaceuticals and biologicals must be kept in mind when strategizing how to apply resources to curing certain diseases. Two of today's major incurable diseases are not directly accessible to biologics—most cancers and neurodegenerative diseases. Biotechnology is not the panacea that the thought leaders said it would be, and the reasons are fairly simple to understand. The deck was stacked against it from the beginning, and it did not require an advanced degree to predict what has transpired. The list of problems associated with biotechnology was all known back in the 1970s, and any project manager would have picked them out and proceeded accordingly. Instead we chose to be lead down a path of academic theory and science worship and are saddled with extremely expensive biologicals that offer limited therapeutic improvements compared to pharmaceuticals. Scores of companies and billions of dollars have produced a handful of extremely expensive products that have had a rather modest effect on disease outcomes. A recent mishap in biotechnology has been an inhaled insulin product developed by Pfizer called Exubera. The company ended up taking a 2.2 billion dollar loss on the product after pulling it off the market 1 year after launching it [4].

It is informative to compare the impact of Doctors Fleming, Marshall, Warren, and Sheskin to the entire biotechnology industry. Three diseases were rapidly cured with innovative maverick insights and no new science. For biotechnology, enormous sums have been spent to generate a huge technology infrastructure and a limited crop of drugs whose cost is straining our healthcare system to the breaking point with no sight of cures on the horizon. The other three diseases were cured using available science, mavericks, and innovation at a cost that was below the radar screen! Bad strategic decisions, or more accurately, no strategic decisions were made over the past four decades, but the problems with industry go beyond that.

START-UP COMPANIES AND WHY THEY FAIL

Biotechnology created such hysteria 40 years ago that it resulted in a stampede of venture capital to many academically oriented basic science companies promising magic bullets. This start-up craze spread throughout the United States with the major centers in the San Francisco Bay area, Boston, and San Diego. Science, which would have been considered too unproven to fund as a commercial venture, began to dominate the biotechnology industry. New terms such as "biopharmaceuticals" were coined to further blur the distinction between pharmaceuticals and biologics so the venture capital (VC) could fund more virgin start-up companies working with small molecules of natural origin.

Virtually all these companies were based on very early stage science with extensive patent protection. They were basically research projects done with investor's money. The large pharmaceutical companies were too smart to spend their own money on these science based start-ups, but they would do corporate partnerships with some of them by licensing rights to the technology; more on that later. Thought leaders became the consultants and advisors of venture capitalists and directed their investments in a manner appropriate to their own interests, virtually all of which the VCs did not begin to understand. But the thought leaders convinced them that cures would be forthcoming if it all worked out. The conflicts of interest in all of this were profound but the thought leaders ruled the day given the complexities of the science. Little did the public or the venture capitalists know just how far-fetched many of these notions were or how much existing knowledge about the difficulties of applying biotechnology to drugs there was. The evidence of the past 40 years tells the real story since the majority of these companies have failed or retreated to a less aggressive business model. Most importantly, virtually no terminal disease cures have emerged as a result of start-up companies.

The initial public offering is the exit strategy of choice for the venture capitalist, and it generally occurs long before a successful product ever materializes. Venture capital, while supposedly well intentioned, has not made significant inroads regarding cures to the major diseases but rather has enriched its investors by other means.

VENTURE CAPITAL: SELLING SCIENCE FOR PROFIT

The role that venture capital has played in medical R&D deserves attention. This industry has been transformed in the past 40 years from a largely unorganized network of angel investors, corporations, private foundations, and investment bankers, to its current state of virtually hundreds of incorporated partnerships that manage billions of dollars of institutional capital. This new breed of venture capitalists, most with big three MBAs, rode the wave of the computer and wireless technology boom that began in the 1970s. These high-technology companies captured the imagination of investors even though, much like biotech, it was difficult for many to understand the nature of the technology or even how they would make money.

The tech and dot-com companies were perfectly suited to the mentality of the then emerging venture capital industry. The investments were largely based on perception rather than reality or past performance. These were companies with stories to tell and not products to sell. Companies were now valued for their potential to generate breakthrough products, and it was common to see market capitalizations in the billions with no hint of revenue on the horizon and huge accumulated losses. Computer-related high-tech companies were the darlings of Wall Street though few of them succeeded. Venture capital firms did remarkably well on the few companies that did make it and even managed to do quite well on the many companies that folded. This was accomplished simply by selling their shares *before* these failed companies began their descent. This new venture capital industry christened a new business paradigm for funding virgin science and mobilized tremendous pools of capital. Success came more often from timing your entry and exit and not from new breakthrough products.

Biotechnology soon became a focus for VCs. The technology seemed to hold tremendous promise since it was different from pharmaceuticals and was ripe with high-powered scientists who promised cures and billion dollar drugs. VCs must generate an exit strategy so they can realize a profit for themselves and their limited partners. Either they identify a buyout by a larger firm or they take the company public at an attractive price and then proceed to sell their shares at the right time. Their primary concern is the exit strategy or how they will cut and run when they have garnered a nice return on their investment. VCs rarely stick with a company until it becomes truly successful in its primary mission. They are often the only investors who escape with a profit or, at the very least, engineer a transaction that minimizes their losses.

A good argument can be made that venture capital has done more to impede innovation than foster it. The enormous amounts of capital they provided to academic companies changed the character of the traditional entrepreneur from one of frugal sacrifice and determination to that of an elite egocentric crusade to showcase glitzy science. The process of cherry picking cures from virgin technologies has not worked. Inherent in the model of venture capital is the total lack of any global strategy for finding cures. Throwing money at hundreds of academic start-up companies with no history of success is extremely inefficient. Most of them have to enter corporate partnerships with large pharmaceutical companies to make the product, fund the clinical trials, and market the product so why bother with starting a company when it just hands off its product to another company? It all seems pretty inefficient, but these are the things that happen when no strategy exists.

PHARMACEUTICALS: WHERE IS THE BEEF?

The pharmaceutical industry has also failed to find cures. The reasons are quite different from that of the start-up biotech companies, but the end result is the same. The mentality of the pharmaceutical industry is much more pragmatic than biotech and biopharmaceutical companies. These folks tend to be much less technology driven and more disease and product focused. Venture capital plays virtually no role in determining what programs get funded, and science tends to be more rationally employed as a tool in drug discovery and development rather than something that must be followed to appease investors or placate academic founders. Pharmaceutical companies are often quick to drop a drug candidate if it looks unsuited to the disease targeted. They are not wedded to technologies for their own sake as are the biotech companies.

Large pharmaceutical companies are revenue driven and spend huge sums generating multiple product opportunities. They have choices when making decisions about which drug candidates to move forward into the very expensive late-stage clinical trials. They have the luxury of prioritizing drug candidates and very often terminating programs when the results are not up to expectations. Decisions are data and market driven. For example, if a pharmaceutical company has a blood pressure drug that is going off patent, it will often develop a variation of that molecule with a slightly better or different side effect and efficacy profile, or a longer duration of action. This will revive the market share of the older, now generic drug but obviously that does little to advance cures. A sizable percentage of what pharmaceutical companies do revolves around this exercise, and it diverts resources from cures. It is quite simply easier for them to make money with these low risk me-too drugs.

New breakthrough drugs are often designated "first-in-class" drugs. These hold the most promise for major improvements in disease outcomes. They generally work via a new mechanism or they act at a new therapeutic target that has not been previously exploited. This increases the chance that the drug might be a cure. But unfortunately a first-in-class drug is much more expensive and risky to develop. The FDA requires much more safety data on these treatments since the drug is acting at a new site in the body, and larger clinical trials are needed to ensure there are no toxic side effects. There is also greater risk that these drugs will incur litigation if patients end up with side effects resulting in death or disability. First-in-class drugs are the holy grail when it comes to cures, but of the new drugs approved each year by the FDA the vast majority is not in this category.

The bottom line is that it is much easier to play it safe, and stick to the "me-too" drugs that do what older drugs do but in a slightly better way. For instance, there are scores of blood pressure drugs that all basically do the same thing but in different ways. The same holds true for diabetes drugs, antibiotics, antidepressants, analgesics, and many more drug classes. The pharmaceutical companies must be incentivized to focus on the high risk first-in-class drugs, and there are ways to provide that as we will see in Part B.

CORPORATE PARTNERING, OIL, AND WATER

When it comes to employing new technologies, a new class of drugs, or a new therapeutic target, large pharmaceutical companies will often partner with the more entrepreneurial start-up companies. This leverages their risk and permits them to try out many of these virgin strategies without making investments in the capital intensive infrastructure required to go it alone. Traditional pharmaceutical companies lack the entrepreneurism of the start-ups, so the partnering of these two different corporate cultures through licensing the start-up technology can inject new directions into their companies, and bolster their patent estates, while giving them a quick way out if the technology fails. For the start-up company, the reward is the credibility of a large established company putting their money and organization behind their embryonic technology. If they are public, their stock price generally doubles in the first minutes after the announcement of a partnership. Obviously, the VCs benefit most from this since the majority of partnerships between pharmaceutical companies and emerging biotech's end up failing and when that happens they are usually long gone.

The partnering phenomenon between pharmaceutical giants and emerging biotech's is an interesting paradox that illustrates the scope of strategic problems that have hindered the discovery of cures. The two corporate cultures could not be more different. One is business and market focused, and the other academic and technology focused. They are at opposite ends of a spectrum that does not include the most important thing we expect from them; therapeutic breakthroughs. The academic biotech's are chasing arcane high-risk technologies whose relevance to practical therapeutics is questionable, and the pharmaceutical firms are too busy tweaking (making me-too or extended release versions, etc.) their biggest selling drugs that are coming off patent so they can drive their sales numbers. Corporate partnering is a gesture for the big pharmaceutical company and a temporary shot in the arm for the start-up company. The success rate for such ventures is therefore low and the process inefficient.

The outlook for the biotech and pharmaceutical industries is not good and will probably stay that way if changes do not occur. The drug industry's problems are not entirely their fault, as we shall see when we discuss the regulatory, legal, and patent establishments. Much of what the industry has had to do to survive has been forced on them by these entities since they have made developing a new drug into a 2 billion dollar process that requires over a decade and exposes the company to constant liability that can cost multiples of the development cost. Until those influences on the industry are reformed the road to cures will be long, something that is tragic given the cost in lives. The industry must be freer to innovate, indemnified to take risks, and incentivized to develop cures rather than focus on treating chronic diseases.

REFERENCES

[1] G. Prindull, Hypothesis: cell plasticity, linking embryonal stem cells to adult stem cell reservoirs and metastatic cancer cells, Experimental Hematology 33 (2005) 738–746.

[2] P.C. Herman, S.L. Huber, T. Herrler, A. Alcher, J.W. Ellwart, M. Guba, C.J. Bruns, C. Heeschen, Distinct populations of cancer stem cells determine tumor growth and metastatic activity in human pancreatic cancer, Cell Stem Cell 1 (2007) 313–323.

[3] B. Fikes, California stem cell agency charts course for future, The San Diego Union Tribune (December 18, 2015) A1–A11.

[4] L. Heinemann, The Failure of Exubera: are we beating a dead horse? Journal of Diabetes Science and Technology 2 (3) (May 2008) 518–529. Retrieved from: https://www.ncbi.nlm.nih.gov/pmc/articles/PMC2769732/.

Chapter 5

The FDA Roadblock to Cures

Chapter Outline

The Food and Drug Administration (FDA) has become a major impediment to the delivery of breakthrough drugs. In 2016 there were only 22 drugs approved. This is absolutely ridiculous given the hundreds of companies spending literally trillions of dollars trying to meet all the whims of what has grown to be an absolutely obstructionist behemoth that has little regard for patients, even when they are faced with death, and no respect for the physician. The FDA is the sole reason for it taking over 10 years and as much as two billion dollars to get a drug approved. This has ramifications throughout the entire healthcare arena from the cost of drugs to the insurance premiums and most importantly to the failure to cure terminal diseases. Regulation of drug safety and efficacy is not a bad thing, but the FDA has taken that function beyond the parameters that make any sense for clinical science given the extreme variability of how the human population responds to drugs.

With the agency's adherence to academic purism, its innate distrust of industry, lack of respect for practicing physicians, and the terminal disease patient's right to life, the FDA repeatedly delays access to new drugs and in doing so creates a negative incentive for the development of cures. All drugs are toxic to some patients under some conditions, and academic purism is fine when it comes to physics or chemistry where the laws of nature are precise but when applied to clinical science it produces diminishing returns. These issues have clouded the FDA's risk–benefit perspective for terminal disease therapeutics, and the cost has been severe both monetarily and in human suffering. There are numerous cases of the FDA's intransigence regarding the approval of terminal

A Roadmap for Curing Cancer, Alzheimer's, and Cardiovascular Disease.
http://dx.doi.org/10.1016/B978-0-12-812796-4.00005-2

disease drugs [1–7], but before discussing that and how this dictatorial and absurd regulation can be fixed it is necessary to have a clear understanding of what the FDA does and how it evolved.

EVOLUTION OF A MONSTER

The mission of the FDA up until the early 1960s was to assure that all foods and drugs sold in the United States were safe. After 1962 that was expanded to include the assurance that a drug was also *effective* in treating the disease it was designed for. The FDA will only permit a drug to be marketed if it meets their safety and efficacy standards. This change greatly increased the time and cost of getting a drug to the public. This also conveyed a great deal of power to regulators not only over the industry but also on a patient's *right to life* when it comes to terminal diseases. Any discussion of curing disease must center on how the FDA is performing its job. Regulatory zeal tends to result in limiting public choices, and unfortunately in the case of the FDA this has essentially doomed cures. Limiting access to a new me-too drug is not a big deal, but for terminal disease it is criminal. What is the FDA, and how does it affect our ability to find cures?

The FDA is made up of divisions based on disease categories. Each is staffed by academicians in chemistry, physiology, pharmacology, and medicine. The division directors, most of whom are medical doctors with little or no experience in the drug industry, wield enormous power and as with most government agencies tend to have lifetime tenure. Division directors control the fate of a drug and rarely are challenged. Contrary to the popular belief perpetuated by the media that industry somehow controls the FDA it is exactly the opposite. There is not a single pharmaceutical or biotech CEO that does not lose many nights' sleep obsessing about the FDA and the fate of their next drug. It is the division directors who deal directly with sponsoring companies, and they can make or break the timeline on a drug as it percolates, and I mean percolate, through the agency. My experience with division directors has convinced me that they often play the biggest single role in moving a drug through the approval process. Get on their bad side and one can expect the road to approval to be substantially more difficult. Companies have little recourse when the FDA takes months, and sometimes *years*, to review their submissions. The companies continue to support their overhead (generally millions of dollars per month) and simply wait for the FDA to give them the go ahead to the next step in the process. Sometimes smaller start-up companies actually run out of capital waiting (my own experience). This is not an issue for the FDA and the companies are helpless responders to its demands and have great difficulty planning their business.

The FDA is understaffed, improperly staffed, underfunded, and often criticized by congress and the media. They have been scolded by congress and activist groups for both being too slow in getting drugs approved and too lax in enforcing safety requirements when side effects occur with approved drugs.

The lack of cures, especially for terminal diseases and the long delays in getting drugs to patients, is in large part the result of a cumbersome and unresponsive regulatory environment and their conservative stance on risk–benefit analysis. Why does it take so long to get drugs approved? Has the FDA created almost insurmountable barriers for drug developers? Is there a way to get lifesaving drugs to patients faster? To answer these questions it is necessary review the steps involved in getting a drug through the FDA.

THE ROCKY ROAD TO DRUG APPROVAL

A new drug or treatment starts out as a vision in someone's mind. Typically, this is a pharmacologist, biochemist, or a physician, but not always. The idea often centers on a biochemical concept, a new therapeutic target, or an innovative insight. This is followed by generating some rudimentary proof of concept data to support the idea and the filing of a patent application: the **first** major milestone in drug development. The cost of developing a medical product is enormous so if one does not own it, there is absolutely no rationale for committing the required capital, so patents in all the major markets are an expensive must.

The drug then has to be synthesized in pure form and tested in the laboratory to establish safety and efficacy in a disease model for which it is intended. This can be accomplished in either test tube type experiments (in vitro) or in animal studies (in vivo). The drug must then be manufactured according to FDA Good Manufacturing Practices guidelines, again an extremely costly task, and shown to be stable with an acceptable shelf life. If these studies are successful, then the preclinical *proof of concept* phase of drug discovery is complete and an Investigational New Drug Application (INDA) is filed with the FDA. The INDA filing and its approval is the **second** major milestone in drug development, and it generally takes about two years to reach this point. The INDA is the passport to human clinical studies and must be preceded by a face-to-face preINDA meeting with the FDA. This is a very important meeting since it is where the company gets its first real feedback from the agency and a sense of whether they are on the right track with their vision for how to get the drug approved by the agency. Who elected the FDA to be the arbiter of this is beyond me because they often do not fully understand the drug as well as the inventors, but that is how the system works. The company outlines its plans for the drug and what clinical studies it proposes to do. Misunderstandings at the preINDA meeting can prove to be very costly for the company and delay or scuttle the entire process.

The INDA documents all the preclinical safety, efficacy, and manufacturing data on the new drug, and the FDA has 30 days to either approve it or ask for revisions that may require more studies on the drug, changes in the manufacturing process or the clinical study protocols proposed, and any number of other issues that the FDA is concerned about. Companies dread being put on "clinical hold." This means they cannot start their Phase I human safety trial until the FDA clarifies something in the INDA. This can take many months, yet often involves a minor clarification. At one of my companies we were put on clinical hold for 6 months over what turned out to be a minor issue requiring no additional work. My clinical and regulatory person told me it is a way for the agency to obey the response time rules when they fall behind. They do this with other response time deadlines as well. I had my hands full trying to explain to shareholders why we had not started our Phase I clinical trial. The FDA has absolute power to hold a drug up for as long as it sees fit. This holds for the biggest, as well as the smallest companies. There is no favoritism in the process; it is data and regulation driven.

The clinical phase of drug development starts with a Phase I safety study where increasing doses of the drug are given to human subjects to determine tolerability. The volunteers get full physical exams after each drug dose and are followed for a period of time to insure that no delayed side effects are present. These studies are the shortest and least expensive part of the clinical trial process, but they are very important since they give a preliminary reading on the safety of the drug at doses higher than those used in later efficacy trials.

On completion of Phase I, the data are submitted to the agency (typically about 6 months to 1 year after INDA's approval), and if approved, they allow the company to begin the Phase II *clinical proof of concept* efficacy trial. This is a critical point in drug development and a positive result constitutes the **third** major milestone in drug development. It also signals an increase in the value of the sponsoring company (if it is a start-up venture funded company) since its successful completion is the first indication that the drug may become a product. The end of a successful Phase II trial often makes a good exit point for some or all of a venture capitalists shares if the company is publically traded. Stock prices generally increase post Phase II. Phase II trials generally take about 2 years to complete.

CLINICAL TRIALS: A VERY INEXACT SCIENCE

A word about clinical trials is warranted before outlining the remaining development path of a drug. Animal studies are done with isogenic strains of rats or mice. This means that the animals are all identical twins and that they will react very similarly to any given drug. The results will therefore be much cleaner than in animals that are not isogenic. In human clinical trials the subjects are not even related, let alone identical twins. This means the results are going to be much more variable. A drug that works in all laboratory animals may only work in some human subjects and not others due to genetic variations. It is important to keep this in mind, especially when terminal diseases are involved. It is also important when considering liability issues. Drug development is not an exact science, and this can result in potentially useful and safe investigational drugs being discarded, something that can be tragic where terminal disease is involved. Humans are the most variable of species, and this is often not factored in to regulatory decisions.

A "scientifically sound" clinical trial needs to satisfy three criteria. Scientifically sound is in quotes for a reason. There are so many uncontrollable variables in clinical research that it borders on inaccurate to call them science, and this must be taken into account when borderline data are evaluated. However, an FDA requirement is that clinical trials must be placebo controlled, double blind, and randomized. Placebo controlled means that there has to be a group of patients who are treated with an inactive facsimile of the drug to determine its true effect. Double blind refers to the fact that the treating physician and the patient do not know who is getting the active drug or the placebo. This controls any bias on the part of both the physician and the patient. Randomization is the process of assigning patients to either the placebo or the active drug group to avoid biasing the trial by assigning less sick patients to the drug group. A computer assigns the drug to each group with a randomization algorithm.

Another important aspect of a "scientifically sound" trial is to set the clinical efficacy endpoints to be measured before the trial starts. Endpoints are things such as blood pressure for hypertension drugs, or blood glucose levels for a diabetes drug, or it could be improvement in the clinical state of the patient. There is always a primary endpoint in a clinical trial, which defines whether the trial is a success or a failure. For Phase I safety trials the primary endpoint is much more general than in efficacy trials and includes such things as global safety, effects on vital signs, adverse effects, such as nausea, pain, etc. In Phase II and III efficacy trials the sponsor has to pick a specific primary endpoint and define to what degree the test drug should affect it for the trial to be clinically relevant. For a diabetes drug, it might be reducing blood sugar levels by 25%, and for a cancer drug, it could be a 30% reduction in tumor mass or a 20% increase in survival time. The fact that a trial can only be considered successful by meeting the primary endpoint is a good example of the academic nature of the FDA. It would seem that if something you were not looking for shows up to be positive and if that were relevant to the patient's best interest it should also be considered a successful outcome. This, however, is not how the FDA looks at it. Picking the right primary endpoint is therefore critical, especially for start-up companies since missing it generally leads to the company going out of business, and the drug is therefore lost even if secondary endpoint results were good. This does not make a lot of sense.

For a Phase II or Phase III trial to be considered successful the primary endpoint must not only be met, but also needs to be accompanied with statistical significance. A given result is considered statistically significant if there is a 5% or less chance that it could have occurred by chance. This somewhat arbitrarily determined criterion is stated as a probability that the effect of the drug was due to chance. This probability is stated as a $P < .05$, and it is considered the proof that the drug actually caused the effect and not a chance-related event. A $P < .07$ means the drug failed and will not be approved, even if it treats a terminal disease. A really strong drug-induced effect will have a P value of $P < .01$ relative to the placebo group.

Often a drug will fail in Phase II or Phase III because the primary endpoint narrowly missed statistical significance, i.e., a statistical significance of $P<06$ or .07 instead of $P<05$. This can be a real problem when the disease in question is terminal and often the FDA just follows the numbers. While this is technically the correct action, when one considers the highly inexact nature of human clinical trials and the fate of the patients in question it seems a silly exercise in academic purism. This is, however, the rule the FDA operates by, and it has created an enormous graveyard of potentially useful drugs.

Phase II proof of concept clinical trials generally involves less than 200 patients at less than 20 clinical centers. Large companies do multiple Phase II trials in different patient groups before deciding on the design of the much larger and more expensive Phase III trials. Small companies do not have that luxury. The large companies also have many different drugs in Phase II, while the small ones do not. The start-ups therefore are much more likely to move into Phase III trials prematurely, while the big guys have the luxury of choice about which candidates they move into Phase III. Is it any wonder that so many small companies fail?

The FDA cares little about the business aspects of drug development. If a company wants to embark on a Phase III trial with less than promising Phase II data that is up to them, as long as the drug has an acceptable side effect profile. This lack of choices in start-up companies wastes huge sums of money and diverts energy and talent from more promising and productive activity.

PHASE III: THE MOTHER OF ALL CLINICAL TRIALS

The **fourth** major milestone of drug development is by far the longest and most expensive: Phase III clinical trials. The FDA requires two pivotal efficacy or registration Phase III trials before a drug can be approved, although they will make rare exceptions for terminal or orphan diseases. These clinical trials often involve a 1000 or more patients and take 3–4 years to complete at a cost of 100 million dollars or more in addition to the cost of keeping the company alive. The only good news is that the odds for success get better since there is already some evidence that the drug works but far more Phase III trials fail than succeed, especially for start-up companies.

It is often stated that 20%–30% of drugs that make it to Phase III will eventually be approved. But this is misleading since it includes both the me-too drugs and the first-in-class drugs. Remember, the me-too drugs are simply variations on a therapeutic theme, such as a new formulation or derivative of an existing blood pressure or diabetes drug. The risk profile for these drugs is much less, and as already discussed, the majority of the drug industry is focused on them. For first-in-class drugs, the success rate in Phase III drops dramatically.

The bottom line is that it takes a long time, a lot of money, and a very high risk of failure to even entertain the prospect of getting a novel, first-in-class drug approved. This is one reason why most companies choose not to venture into the realm of breakthrough drugs for terminal diseases. It is just too difficult to get anything approved, and if you are lucky enough to make it to approval, your patents are probably running out by that time. Who would be dumb enough to want to do this; is it any wonder why cures are not happening?

Assuming that both Phase III trials go well and the primary endpoints are met, the sponsor now reaches the **fifth** major milestone of drug development—the filing of a *New Drug Application* (NDA). The NDA takes 6–12 months to prepare and submit to the agency since it contains all the information on the drug in its now 7–10-year journey through the development process. The FDA has 6 months to review the NDA but often takes longer since they frequently want additional data or clarification before approval.

The penultimate **sixth** step to approval of the NDA is the FDA advisory committee meeting. This is an assemblage of distinguished outside thought leader physicians and scientists charged with reviewing the NDA and voting yes or no on the safety and efficacy of the drug. The members of the advisory committee, usually about 12–15 people, are typically academics with little clinical experience treating patients and even less experience in the drug development business. Thus they often lack the real-life frontline contact with the patient and the family dynamics that surround the terminal disease experience, and this can affect how they evaluate the risk–benefit issue. The FDA selects the members of the advisory committee. The recommendations of the advisory committee are generally followed by the FDA but not always. The agency has an obligation to advise the company within 6 months whether the NDA is approvable or not, but in reality, the NDA approval, the **seventh** step, generally takes more than 1 year since the agency often has conditions or comments regarding the labeling and product insert materials. When one does the math this brings the approval timeline to about *ten to twelve years* for a typical drug.

This is one monster of a product development cycle. The patent life is usually more than half over, and the science behind the drug may have changed in the interim. The drug industry is not for the faint of heart. The regulatory path is so formidable that one wonders why any business entity would be foolish enough to engage in such a high risk, cost-intensive activity. Most of the large companies deal with this situation by carefully choosing their product targets to improve the odds of successfully navigating the arduous product approval process. Me-too drugs look very good in this light. The smaller companies generally are more aggressive about venturing forth with first-in-class drugs, but they rarely succeed since they often run out of capital or the drug just does not work. The scientific precision sought by the regulators and their academic advisors is in theory desirable, but with terminal disease drugs it makes much less sense and leads the industry to me-too drugs. Clinical trials are not an exact science, yet the regulators treat them as if they are since their benchmarks and decisions are purely academically driven. Both the patient and the physician are on the outside looking in, and all too often must wait until it is too late before the new drug he or she may need is officially approved by the regulators. There has to be a faster way to move forward.

MUST WE BE SO TIMID FOR ALL DRUGS?

For me-too drugs that offer incremental advances over existing drugs, the current conservative regulatory landscape is barely tolerable, but when it comes to cancer and Alzheimer's disease it is at best maddening and at worst criminal. It infringes the rights of patients and the physician–patient relationship. There needs to be a different set of metrics and a more realistic appraisal of the patient's right to life when terminal diseases are involved. While the FDA has made an effort to address this problem, there have been feeble attempts at modifying the current academically driven system of drug approval, rather than the more fundamental changes that are required.

There is currently an accelerated review process for drugs that treat life-threatening diseases for which there is not adequate therapy. These provisions include "expedited review" and "fast-track" status. This status provides for faster response times from the FDA to sponsor initiated communications and, at least in principle, one Phase III trial to gain conditional approval. While this legislation, which was passed in response to AIDS activists and did accelerate, AIDS drug approvals, it has over time been eroded and the agency has settled back into its traditionally conservative pattern. It is now only rarely permitted by the agency, and this has cost many lives. Numerous new drugs have been held up by the FDA rather than expedited with perhaps the best example being the prostate cancer drug Provenge [8]. This was in spite of no serious adverse safety issues experienced with these drugs. What is more, the program as envisioned does not go far enough as will be discussed in Chapter 8. Like so many bureaucratic fixes it was simply a variation on a flawed theme rather than the quantum change that is required.

TERMINAL DISEASE PATIENTS CONDEMNED TO DEATH

Society condones a patient's right to refuse palliative life-extending treatments when they become terminally ill, yet it withholds from that same patient the right to take a drug that may save his or her life just because it has yet to meet an arbitrary and overly conservative regulatory metric. We therefore provide the right to choose death but not life so does it make sense to even consider safety in this situation? The current regulatory climate has been aided by an aggressive tort system, which eviscerates drug companies for even the most nebulous of drug adverse events. This paralyzes the development process and makes the FDA more cautious than it should be with regard to terminal disease drugs. It also ignores the physician–patient relationship by limiting treatment options.

Cancer drugs, such as Gleevec for leukemia, Eloxatin and Erbitux for advanced colorectal cancer, Tarceva for lung cancer, Provenge for advanced prostate cancer, and Bexar for lymphoma, are examples of drugs that were, for one reason or another, held up by the agency and not gotten to patients soon enough. There are many more examples, all of which are a matter of public record, and all did eventually gain approval. Many of these were approved much more rapidly in other countries where the safety record is not substantially different from that in the United States. The practice of accelerated review in selected diseases is not sufficient. There must be a policy of accelerated *availability* to those patients who have exhausted all other treatment options. The current "right-to-try" initiative [9,10], which would grant terminal disease patients accessibility to investigational drugs, is a start; but it is not clear whether these state initiatives will work since federal laws will have to be considered and the FDA may prevail. There is, however, a Bill in the Senate pending [10]. The FDA will likely have ways to get around this, but it is a step in the right direction.

The strict academic mentality of the FDA regarding risk–benefit scenarios is especially maddening when one factors in the imprecise nature of clinical trials. A Phase III trial may involve over a 100 trial sites with only 10 or 20 patients per site. These can include the entire United States as well as many other countries. Medicine is practiced differently at these sites, and the patients are different with respect to a host of genetic, environmental and cultural variables. Drugs will react differently in these patients, and this will complicate the study results. Is a $P<.05$ the proper metric or should it be $P<.10$? How many potentially useful drugs have been missed in this highly inefficient process? Yet we continue to deny access to a potential lifesaving drug simply because it narrowly missed a statistical endpoint in a Phase III trial. It brings to mind the T.S. Elliot quote, "Where is the wisdom we lost in knowledge?"

THE MISTAKE THAT CREATED THE MONSTER

The regulatory mentality practiced in the United States is hostile to innovation and risk intolerant to the point of ineffectiveness. This is largely the result of overreaction to a small number of drugs that have, during the past 50 years, proven to be unsafe. It turns out that the response to these incidents has proven far worse than the initial crisis for which the remedies were designed. A look at the major crisis that lead to this FDA monster that now shackles terminal disease drug development illustrates the point well.

The thalidomide affair is a distant memory, and the real history of what transpired was never adequately considered when the FDA was revamped in its wake. The FDA did its job in identifying thalidomide as an unsafe drug (discussed earlier) since it was never approved in the United States, yet its effect on our regulatory climate has resulted in a virtual paralysis of new drug development by the FDA.

Drugs are foreign substances, and they are, by definition, all unsafe for certain people under certain circumstances. The real issue is the rectification of the *risk profile of a drug* in the context of its therapeutic benefit. If a drug saves a

100 lives for every one person it harms, then it needs to be used until something better is available. The FDA has assumed the role of the grand arbitrator of which drug will be available, and this works when other treatment options are available. The word terminal seems not to matter to the FDA.

Government often passes new laws as a knee–jerk reaction when things go wrong. In the thalidomide case the FDA did nothing wrong since it never approved the drug. In spite of that this episode resulted in new mandates for the FDA to establish efficacy for all drug approvals. A good case can be made that this decision ended orders of magnitude more lives compared to having just left the FDA mandate to establishing safety for drug approval. In this hostile regulatory environment the industry has no choice other than play it safe with me-too drugs where the risk of unknown side effect profiles is less likely to trigger the heavy hand of product killing regulatory conservatism and an overzealous legal system.

The FDA is in every sense the eight-hundred pound gorilla in the drug development business. While its mission is reasonable, it has evolved into an entity that inspires fear in virtually every drug company. It holds absolute power over our healthcare system, and especially, the quest for cures. Cures involve innovation, risk, and judgment, but the FDA often punts in situations where a clinical trial result is anything short of perfect. It takes the safe route and demands another trial, which takes years. So patients with no hope have to wait and patiently die.

Patient advocacy groups such as the Abigail foundation [11] and the right-to-try movement, to name a few, have pleaded with the FDA in an effort to make promising investigational drugs available to terminal disease patients. These efforts are rarely acknowledged and end up being negated by the whim of the FDA. Others have pleaded for shorter development paths and increased funding of the agency [12,13] but have had little effect. The FDA clearly needs to change its tune in fundamental ways and start facilitating cures rather than stymie them.

SURGERY, AN EXAMPLE OF WHAT FEWER REGULATIONS CAN PRODUCE

Perhaps the reason that surgery has progressed so much further than the pharmaceutical and biotech sectors is the fact that surgical procedures are not subject to stringent FDA regulation. It is common to see mortality rates of several percent or more in various major surgeries yet the surgeon is still able to perform these operations since they assess the risk benefit themselves and make decisions with their patients on whether the risk justifies the reward without FDA intervention. Coronary artery bypass surgery is one such example. Several hundred thousand of these surgeries are performed every year in the United States alone. While the decision between *certain* death and a promising investigational drug that may work is prohibited, why is this high-risk decision (relative to drug therapy) left to the patient and the surgeon and rarely challenged by FDA intervention?

Advances in surgery have been impressive because surgeons are, relatively speaking, not micromanaged as are drug developers. Organ transplants, noninvasive procedures, robotic surgery, laser surgery, and microsurgery are just a few of the advances. Surely much of this would have not happened if the FDA was more involved. The surgeon and the patient were quite capable to handle things on their own.

REGULATION AND THE PATIENT–PHYSICIAN RELATIONSHIP

Regulators by definition establish a barrier between the consumer and the service provider. They also have a tendency of becoming omnipotent as time goes by. This has definitely happened with the FDA and it is a major roadblock to not only curing terminal disease but also to treating it. Not only has a terminal disease patient's right to life been sequestered but the physician's duty to the terminal disease patient has also been usurped by the FDA. When a promising investigational drug appears on the landscape the terminal disease patient should not have to wait an inordinate amount of time to try it because of a regulatory agency. Under the guidance of the treating oncologist, cardiologist, or neurologist it should be possible for the terminal disease patient to try the investigational drug. The prognosis without the drug is certain death so what could be worse.

The drug companies can get nervous about their investigational drug being used outside of a clinical trial setting partly because of liability if the drug is not used exactly as the trial protocol dictates or if adverse events occur, but these issues can be dealt with through the proper informed consent procedures and documentation. The investigational drug would also have had to be shown in a Phase II clinical trial to be free of major side effects for this approach to be used and the drug companies indemnified from legal prosecution.

The FDA is doing more harm than good in its current state [3–8] when it comes to terminal disease drugs. Safety is clearly a critical function as is some evidence of efficacy but the agency has taken the later metric to extremes and retired many drugs to oblivion, while delaying others for years while patients suffer. The situation becomes ridiculous in the context of terminal disease patients, and it has to change. It is not enough to reform the FDA, it must be fundamentally rethought and become a partner with industry to facilitate drug discovery and availability to terminal disease patients rather than impede it.

REFERENCES

[1] J. Whalen, Hurdles multiply for latest drugs, The Wall Street Journal (August 1, 2011). Retrieved from: www.wsj.com/articles/SB10001424053111904233404576459851152423110.

[2] A. Von Eschenbach, Toward a 21st-century FDA, The Wall Street Journal (April 15, 2012). Retrieved from: www.wsj.com/articles/SB10001424052702303815404577331673917964962.

[3] R. Goldberg, Government is stifling medical innovation, The San Diego Union Tribune (March 13, 2011). Retrieved from: www.sandiegouniontribune.com/opinion/commentary/sdut-government-is-stifling-medical-innovation-2011mar13-story.html.

[4] The FDA vs. Austin Leclaire(Editorial), The Wall Street Journal (April 22, 2016). Retrieved from: www.wsj.com/articles/the-fda-vs-austin-leclaire-1461281386.

[5] The real FDA scandal(Editorial), The Wall Street Journal (February 6, 2008) A18.

[6] The FDA and slower cures(Editorial), The Wall Street Journal (February 28, 2011) A18.

[7] Where's the drug, FDA?(Editorial), The Wall Street Journal (July 2, 2016). Retrieved from: www.wsj.com/articles/wheres-the-drug-fda-1467413266.

[8] W. Faloon, FDA delay of one drug causes 82,000 lost life-years, Life Extension Magazine (November 2010). Retrieved from: http://www.lifeextension.com/magazine/2010/11/fda-delay-of-one-drug-causes-lost-life-years/Page-01.

[9] B.A. Cohen-Kurzrock, P.R. Cohen, R. Kurzrock, Health policy: the right to try is embodied in the right to die, Nature Reviews Clinical Oncology 13 (May 24, 2016) 399–400, http://dx.doi.org/10.1038/nrclinonc.2016.73.

[10] R. Nelson, "Right to Try" bill in Senate for terminally ill patients, Medscape Magazine (November 20, 2016). Retrieved from: http://www.medscape.com/viewarticle/863336.

[11] The Abigail Alliance. Retrieved from: http://www.abigail-alliance.org/story.php#.

[12] A. Von Eschenbach, Medical innovation: how the US can retain its lead, The Wall Street Journal (February 14, 2012). Retrieved from: http://www.wsj.com/articles/SB10001424052970203646004577215403399350874.

[13] FDA should encourage, not delay, medical innovation(Letter to the editor), The Wall Street Journal (February 24, 2012). Retrieved from: http://www.wsj.com/articles/SB10001424052970204792404577229641193886650.

Chapter 6

The Liability Barrier

Chapter Outline

Medical malpractice, drug side effects, drug clinical trial failures are all a litigator's dream come true, and they constitute a major impediment to the entire medical R&D process. While it is relatively easy to prove that a physician may have committed criminal negligence if he or she was derelict in dispensing appropriate medical treatment, it is often very difficult to prove that a drug caused harm to a patient, especially when the incidence of such harm is very low and the benefit of the drug is factored in. Drugs are foreign substances and humans are the most variable of species. We intermarry with other races from different continents and our gene pool is in a constant state of change. The result is that when you mix drugs and the human species some unpredictable things can happen. We have already discussed this as it relates to the variability of clinical trials. This nebulous aspect of how drugs affect different people is compounded when disease is added to the equation. All this culminates to a very attractive breeding ground for litigation and has spawned a lucrative subspecialty of the legal profession, one that has exploded over the past several decades. Unfortunately the rise in medical malpractice and drug side effect litigation has cast a chill over the entire drug development arena in the United States and has forced the entire system into a very conservative and timid posture, which is not conducive to finding cures.

EVERYONE RESPONDS TO DRUGS DIFFERENTLY

Prescription drugs often react differently across populations. A physician will often start a treatment course with a low dose of drug to assess the patient's reaction. Some people metabolize the drug more slowly than others, which results in higher levels of the drug and exacerbate the drug's effects, both good

A Roadmap for Curing Cancer, Alzheimer's, and Cardiovascular Disease.
http://dx.doi.org/10.1016/B978-0-12-812796-4.00006-4

and bad. Others could be fast metabolizers and not be affected by the drug at all. The physician and the patient need to work closely to find the right dosing regimen. One size does not fit all when it comes to drugs.

Some foods inhibit drug metabolism, which can result in enhanced side effects, certain human genotypes have effects on a drug's actions, drugs can interact with each other, various herbal preparations and nutraceuticals can affect drug action, and certain people are allergic to specific drugs. Herbals are particularly problematic since they vary so much from batch to batch, and it is impossible to know what is in them. Patients and physicians cannot be expected to monitor all of these variables, but they try. This is why clinical trials are so difficult. All of these factors add enormous variation to both the therapeutic and side effects of drugs. Add to this the fact that many drugs are used "off label" to treat diseases for which they have not been specifically tested. While this can be a good thing for the patient, the litigators see it as an opportunity when things do not work out as hoped.

The public often does not realize all the factors involved in drug therapy, but the industry and the Food and Drug Administration (FDA) must try to account for all of them. It is not an easy task. If one wants zero-risk drugs, then there would be no drugs at all. The issue becomes what degree of risk is acceptable for which therapeutic benefits? This is a judgment call that the FDA makes when it approves a drug. Balancing the therapeutic need with the potential risk is not an exact science. Cancer chemotherapy drugs have horrendous side effects. They are in fact cytotoxins. They not only kill cancer cells but they also kill normal cells throughout the entire body but at a slower rate! They gain approval because the therapeutic benefit is life rather than death for some period of time.

Approval of a drug is, as we have seen, perhaps one of the longest, most expensive processes imaginable. Both the FDA and the sponsoring company work to make sure the new drug is extensively tested and that all the data are closely scrutinized. Nothing is hidden; every adverse event associated with the drug during clinical trials is immediately reported and thoroughly investigated. Failure of a company to disclose all the data on a new drug to the FDA results in very serious actions and few companies violate that code. In the United States, the case most often made about the risk–benefit mentality is that the FDA is too *intolerant* of risk. This is one reason why it takes much longer to get drugs approved in the United States. This conservative attitude about risk–benefit profile criteria for a new drug is in large measure fostered by the litigious legal system we have created.

The FDA can mandate that a drug carry a "black box" warning on the label if there are serious potential side effects in selected patients. This warns both the physician and the patient to be on the lookout for these potential side effects and is generally the last resort before actually pulling a drug off the market. Black box drugs generally treat more serious diseases for which there are few other drugs available. It is the FDA's attempt to keep lifesaving drugs with more serious side effects available to patients, but it often counts for little in the emotionally charged courtroom environment.

Drug therapy is therefore a mine field for both drug companies and regulators when liability is not clearly defined. The safest drugs can be harmful to some, yet lifesaving for many more people. Sometimes the incidence of toxic effects will be so rare that they will only show up when millions of patients have been treated. These types of situations push statistical methods to limits and reduce their effectiveness or make them useless. In almost every case, when one weighs the benefit of an approved drug with regard to the risk, it is apparent that the drug is doing much more good than harm. Unfortunately for the drug company, it is the harm that almost usually wins the day and ends up costing the drug company billions of dollars. There is no clear definition of liability for the drug company, even when the FDA has approved the drug in question. The FDA is immune from liability. It is easy to see why drug companies often chose to go after the low-hanging fruit when it comes to developing drugs. The risk of litigation for me-too drugs is much lower than that for the first-in-class drugs that have the best chance of being cures.

WHERE DOES LIABILITY BEGIN?

Clinical science is not precise enough to even know all the variables involved in human physiology. A person may react one way at 20 years of age and entirely different at 35. Some patients are more prone to becoming sensitized or desensitized to a drug with consequences that can be significant. It is very difficult in this landscape to assign cause and effect with any precision. Often the side effects in question were observed during the clinical trials that supported approval but considered manageable by regulatory authorities. Where does liability begin in a situation where an impartial and very stringent regulatory authority vouches for the safety and efficacy of a drug? This is the issue, and under the current system it is not clearly defined. It is left to individual juries who often do not appreciate all the variables involved in how drugs affect different people and rather defer to the emotional aspects of the situation.

The real world of drug use is much different than the clinical trial setting. Everything is carefully controlled in clinical trials. The patient groups are age and sex matched, their disease state must meet strict criteria, and their medical care is closely monitored. In the real world, the patient is much less controlled; no one is watching his or her behavior and recording the medical response to the drug or even assuring that the patient is following the dosing correctly. Often, the treatment is in progress months or years before the side effects are noticed. There is no relevant data on the plaintiff's compliance on issues such as adherence to the dosing regimen or other relevant behavior. It is therefore difficult, if not impossible, to establish clear cause–effect relationships in this sea of uncertainty regarding the actual usage of the drug and how stringently the patient and physician followed the manufacturer's directions. Might the patient have taken the drug with alcohol or after taking another drug that interacts with the drug in question? No one knows.

Let us assume that a cause–effect relationship is demonstrated between a drug and a serious adverse event for a particular patient. The relevant question then should be how many patients have benefited from the drug? If the harm done was a death, then one needs to ask how many patient lives were saved by the drug. Let us consider the surgery analogy again. There is a significant mortality rate associated with all major surgeries. When a patient dies on the operating table there is no knee–jerk reaction to sue and ban that particular surgical procedure. Clearly, the surgery killed the patient. In the case of the drug it is often far from clear. In the surgery situation, the expectation of risk is built in, whereas with drugs it is not, even though the product inserts for all prescription drugs read like a death sentence. There are so many warnings that most people who read them have to wonder whether they should even take the drug. The risks are clearly stated while the benefits are only alluded to. This is especially apparent in the TV commercials for prescription drugs. One wonders why the drug companies even bother because 90% of the advertisement is spent on the potential toxic effects of the drug. The risks of drug therapy are literally pounded into the consumer. Surely that, coupled with FDA approval, should provide some protection from liability.

In drug litigation, the defendant is a company with deep pockets, and that is what motivates plaintiffs and their attorneys. Companies are the evil Goliath, while the patient is David. Often the only real beneficiaries of drug safety litigation are the attorneys. For every plaintiff that has a decent case there are many more that are just riding the out of court settlement train. It is informative to look at some recent cases where drugs have been the subject of market withdrawals or lawsuits because of side effects.

MEDICAL CLASS ACTION LAWSUITS: A LITIGATOR'S DREAM

Vioxx was approved in 1999 as an antiinflammatory drug for the treatment of arthritis pain. The therapeutic target for Vioxx is the same as that for aspirin with the only difference being that it is more selective than aspirin. Aspirin inhibits a group of enzymes that block inflammation and pain. These enzymes are called cyclo-oxygenases (COX), and there are two major types. Aspirin inhibits both COX-1 and COX-2 enzymes. Vioxx selectively blocks COX-2, and has the same benefits as aspirin, but lacks some of aspirin's side effects, which are mediated by COX-1. These side effects are stomach upset and the inhibition of blood clotting. The latter side effect can be life-threatening since the inhibition of blood clotting can cause excessive bleeding and death in a substantial number of peptic ulcer patients. It is estimated that over 20,000 lives per year are saved by using Vioxx when compared with aspirin-type drugs. Other variants (me-too drugs) of selective COX-2 inhibitors such as Celebrex and Bextra are marketed by other drug companies. Up until 2004, Vioxx had been used in over 80 million patients in scores of countries, and the COX-2 inhibitors were established as a very useful new class of pain reliever. Vioxx was also being tested for other diseases, including colon cancer recurrence, and that is where some potential problems started to appear for the drug [1–3].

In 2004, the results of a large multiyear study conducted in colon cancer patients to determine if Vioxx treatment would prevent the recurrence of cancerous polyps in the large bowel were released. It was in that study and in that group of patients who had already had colon cancer that an increased risk of heart attacks was noticed in the Vioxx group. The increased risk of heart attack was only detected after 18 months of treatment with the drug, a dosing regimen not evaluated in previous clinical trials for inflammation and pain. The drug company sponsor immediately and voluntarily withdrew the drug from the market in September, 2004. The US FDA and Canadian regulatory authorities never withdrew *approval* of the drug since both felt that the risk benefit of the drug was still favorable with the dosing regimen specified in the product insert for pain and inflammation, which was far less than 18 months. The FDA did suggest additional labeling warnings but the company still did not reintroduce the drug. So, the company took immediate action without prompting from the regulatory authorities. Many thought the company was overreacting, myself included.

In spite of this seemingly responsible action by the drug company, the attorneys sprang into action and began trolling for clients. The drug company originally decided to fight each suit individually and was successful in about half of the cases. In 2005 the New England Journal of Medicine (NEJM) published an editorial charging that the company withheld data from a previous trial regarding three patients who suffered heart attacks in the Vioxx group [4]. The drug company denied this claim that there were study design–related reasons to delete those patients from the data. The FDA apparently agreed with the company since it never reprimanded them. This unsubstantiated allegation led to a sharp increase in lawsuits and a sharp drop in the company's market value. They were forced to lay off 7000 workers and close 8 research and production facilities.

After 3 years of individual litigation where the company won more cases than it lost, and about two billion dollars in legal and settlement costs the company struck a deal with the plaintiff's attorney to settle all the potential liability for 4.85 billion dollars. The company did not admit guilt regarding withholding information on the drug or willfully exposing patients to harm. The settlement was booked as a cost of doing business, and the total cost to the company was over $7 billon.

While there was reason to be concerned about Vioxx after the colon cancer trial was announced, one has to question the validity of the drug company liability given the facts of the situation and the continued support of the drug by the FDA. Numerous patients and their physicians were without this drug as a treatment alternative for patients that have bleeding problems and therefore need the drug to prevent fatal blood loss.

Again, all drugs have side effects. With Vioxx they happened to appear after 18 or more months of treatment in colon cancer patients so they were rather difficult to detect since the clinical trials required for the pain indication involved much shorter treatment periods. In spite of that, many of the lawsuits were filed by patients that had only been taking the drug for weeks or months. No one seemed to care since the mindset is that the company is the bad guy and they will settle for purely business reasons. Clearly, the typical feeding frenzy mentality was at work here—let us see what we can get if we milk the system like everyone else is.

The seven billion dollars spent on litigation and judgment costs could have been better spent on research. There was never any malicious intent established and no way of knowing what the full side effect profile of the drug was until it was in the real-world setting and being given for 18 months or more to a unique patient group. In fact, colon cancer patients may have been atypical in this regard. No one really knows if cancer-free patients would have increased heart attack risk, but the litigators behaved as if it was so and extrapolated the results of the colon cancer trial to the arthritis patients who were their clients. If one assumes that it costs about a billion dollars to bring a new drug to approval; that would translate to seven new drugs for this one company alone.

Avandia is an oral diabetes drug that effectively controls blood glucose levels in Type II diabetes patients and is a first-in-class drug that increases the effectiveness of insulin in getting glucose into cells, a key problem for diabetics. It was among the first of the insulin sensitizers and offers not only a new and exciting way to treat diabetes, but may also be important in other diseases related to diabetes, such as Alzheimer disease, since the two diseases often appear together. It offers an important alternative to other oral antidiabetic drugs that work via different mechanisms and can have serious liver side effects. The complications of diabetes are life-threatening and the more effectively one manages the disease the less likely the complications become. Preliminary clinical trials also indicate that it may be effective in slowing cognitive degeneration in a certain subgroup of Alzheimer's patients [13].

The assault on Avandia has many similarities to that of Vioxx. Lawyers see an opportunity and swoop in to over ride the company and the FDA. The FDA is immune from prosecution but the company is not only fair, but in fact, favored game. Academic thought leaders sometimes facilitate the legal attacks and lend credibility but more often than not the physician and the patient are not well served by the outcome. We have discussed this deep schism between academia and industry previously and the Avandia case further demonstrates just how deep the antagonism is and how damaging it can be to the advancement of healthcare. Some 6 years after being approved in over 25 countries, including the United States, a study was published in, again, the NEJM. The study was a metaanalysis of 42 published literature studies involving Avandia [5]. What is most revealing about this study is the manner in which it was presented to the media and the effect it had on both patients and their physicians.

Metaanalysis is a technique in which the data from many literature studies are pooled to attain larger data sets to determine if any new properties of a drug can be gleaned. It is much less accurate than a large single study since the individual studies that constitute a metaanalysis are each designed to look at different things in different patients. Also, published academic studies are typically not done according to the Good Clinical Practice (GCP) guidelines that the FDA demands from companies. GCP clinical trials are extremely expensive since extensive patient data are required, and it all must be documented on case report forms. Academic investigators cannot afford to do studies in this manner. Non GCP studies are very difficult to audit and have much less veracity regarding quality control and patient information. Academic trials are also done at fewer sites and are smaller in size so it is more likely that the patient populations will be less representative of real-world medical practice. The FDA does not accept metaanalysis data in support of drug approvals. For these reasons, it is generally accepted in clinical research that it is entirely improper to draw hard clinical conclusions from metaanalysis data.

The NEJM metaanalysis study of Avandia showed a small increase in cardiovascular complications, such as heart attacks and congestive heart failure, in the patients treated with Avandia. Congestive heart failure was already listed in the product insert as a rare side effect for Avandia, so this was not a new finding. In fact the FDA was also looking at the same data with its own metaanalysis of the same studies. The company sponsor of the drug did a similar metaanalysis several years earlier and followed it with a huge single study trial that showed no serious cardiovascular side effects associated with Avandia. The issue of potential cardiovascular side effects of Avandia was therefore under careful scrutiny by both the FDA and the sponsoring company, and the NEJM study investigators knew this.

In that landscape, the authors of the NEJM study and the journal editors should have at least waited for the FDA to finish their analysis or consult with them before causing a panic by releasing their findings. Instead they chose to issue a press release on their findings, which resulted in a damaging round of media hype and a flood of patient, physician, and, you guessed it, ambulance chaser activity. The litigation genie was out of the bottle and now litigators had websites up and TV commercials trolling the universe for potential plaintiffs. The sponsoring company and the FDA were assumed to be guilty in spite of their both being on top of the issue and in possession of superior data to that which constituted the NEJM report. Physicians who relied on the NEJM were faced with deciding to continue or stop treatment with Avandia. Patients were frightened, the FDA and the drug company were traumatized, and Congress was holding hearings. The end result was much the same as for the Vioxx case several years before. Avandia was not pulled off the market by the FDA. It turned out that while metaanalysis studies did show a small risk of cardiovascular side effects, the more powerful individual studies designed to show such effects were unconvincing and determined that no significant effect was noted [6–8].

The same litigation attorneys that had mined the Vioxx situation several years before were planning Avandia class action lawsuits within hours of the

NEJM announcement. The facts of the case were irrelevant, it was feeding time. There was no consideration given to the reality of the situation and the system was in place to indict the entire evil drug development establishment; the company, the FDA, and the physicians prescribing the drug. The media, academia, and the tort system are extremely powerful forces, and when they join in as a group, nothing is off limits. The final result was a huge class action suit, which was settled by the company followed by the FDA clearing the company and dropping the sales restrictions for the drug. All this activity and grief and in the end after all the damage was done it was proven to be for naught. The company was not compensated, the FDA wasted a lot of precious time, and the academics got off scot free [9].

Tysabri is yet another example of a drug that was victim to a very rare side effect. This immunosuppressive drug was approved to treat multiple sclerosis patients and widely hailed as the most effective treatment available. Tysabri was designated as a "disease modifying" drug for MS. This is one step short of a cure since it indicates that the drug substantially changes, for the better, the patient's life. About a year after approval it was noticed that two of the approximately 25,000 patients who received the drug were diagnosed with a rare form of brain white matter tissue infection called progressive multifocal leukoencephalopathy (PML). The response was swift as multiple law firms filed actions. This time both the lawyers and the FDA stepped in and the drug was recalled. A key fact of this case was that the two patients who were diagnosed with the rare brain infection were unique in that they were taking Tysabri and a second immunosuppressive drug called Copaxone.

The most likely explanation of the two brain infections was that the patients' immune systems were so severely depressed by the combination of both drugs that they became susceptible to this very rare brain infection PML. A more appropriate response for a disease modifying drug of this magnitude might have been to modify the label to *prohibit* patients on Copaxone or other immunosuppressants from taking the drug and to restrict it to only the sickest MS patients until the issue could be resolved. If the litigation environment were more reasonable, this would probably have been the course of action. Tysabri is an MS patient's last resort, yet for almost a year it was not available [10].

It turns out that Tysabri was reintroduced to the market with the blessing of the FDA in less than a year with a modified label when it was clearly determined that the drug should only be used in the absence of other immunosuppressive drugs and with a strict follow-up regimen to guard against PML. In spite of these lawsuits are still filed against the company to this day more than 10 years after the fact. As ridiculous as this may seem it clearly makes the point of just how senseless the current legal system is and the need to reign in this monster.

Silicone breast implants constitute yet another classic example of how the tort system can irrationally and without any evidence other than an irate plaintiff, traumatize the practice of medicine, and destroy companies, and waste enormous capital. Silicone implants were first used in the 1960s and were not regulated by the FDA since they were considered "surgical devices" and not drugs. In 1976 this authority was given to the FDA by The Medical Devices Amendment to the Federal Food, Drug, and Cosmetic Act.

In 1977 a lawsuit was filed against Dow Corning [11,12], the manufacturer of silicone implants, by a Cleveland woman claiming she endured pain and suffering when an implant ruptured. Dow Corning lost the case. This was followed by other cases, all of which claimed that different medical conditions were associated with silicone implants. In the early 1980s a private citizen's action group declared that silicone breast implants caused cancer, and this caused a lawsuit avalanche. Claims of everything from fibromyalgia to autoimmune disease were all of a sudden being attributed to the implants. In the early 1990s the FDA restricted use of silicone implants until safety data could be evaluated. Dow Corning continued to be bombarded with increasingly large lawsuits based on unsubstantiated anecdotal data. It was open season for the lawyers and plaintiffs, and Dow Corning was giving out money.

There have been many studies since the 1980s and no connection has ever been proven between any disease and silicone exposure. But still, the lawsuits kept coming by the thousands, and Dow Corning was forced into bankruptcy from which it took 9 years to emerge! The manufacturer was declared guilty by the media and various public action groups who were increasingly successful in convincing juries that silicone caused a myriad of diseases in spite of the numerous studies showing just the opposite [11,12].

Silicone breast implants were eventually absolved of any toxic effects and reintroduced to the market. Absolutely nothing was accomplished by the enormous litigation. Not one plaintiff was protected from anything, and not one dime was ever returned to the company. While there are examples of medical litigation that are justified, there are far more that can only be characterized as theatrical passion plays of pure greed and opportunism where lawyers tempt the public to milk the system. A means to an end for both the plaintiff and the litigator that does not begin to understand the nature of drug therapy and the risk–benefit scenarios at work. The irony is that the plaintiffs and their attorneys appeared to be greedier and more opportunistic than the companies they professed to be punishing for just those reasons.

I could go on with more examples of what I have observed over the years. What is it that fuels such folly in our medical legal system? Is the legal profession a pure instrument of exploitation and excess? The extremely low threshold for legal action and the enormous awards that brilliant litigators secure for their clients have made the business of drug-related class action lawsuits irresistible. No consideration is given to the real-world risk–benefit equation that is the essence of clinical medicine. Every intervention into the body of a living organism is a risk. One does not dismiss a drug as dangerous simply because it has adverse effects in some patients, especially when the patient is assumed to have been totally compliant in their use of the drug. There is a complex interplay between risks to a few and the good of many. Some risks are harder to identify, especially when they only appear after extended periods of treatment or the sponsor ventures into another disease state than the drug was initially approved for. This is unfortunate because it acts as a disincentive for companies to broaden the scope of a drug's uses. Often the patient group is prone to the risk in question. For example, in arthritis patients there is a higher risk of cardiovascular problems such as those observed with Vioxx. These patients are older, sedentary, and often overweight. The cause–effect relationship can be exaggerated in such groups.

SHOULD WE TRUST THE DRUG COMPANIES?

Why is such a flawed medical tort system allowed to exist? Perhaps the answer can be found in the distrust of business in general that the American public has developed over the years. With regard to medical and drug lawsuits this attitude might find its roots in the tobacco companies. For decades these companies fought the well-proven fact that smoking was linked to lung cancer, an obviously erroneous position. The public and the media realize this and have transferred that distrust of tobacco companies to drug companies. The logic here is flawed since tobacco is not regulated and tobacco companies are not in the business of treating disease. This is not a fair comparison to the drug companies that are subject to extreme regulation and are spending billions to try and treat disease.

Tobacco companies have no resemblance to drug companies. They deal in a commodity that is within the law to sell. The resistance of tobacco companies in facing up to the health risks associated with their product has no relationship at all to drug companies and their product development cycles. Cigarettes and alcoholic products are what they are, and the risks are common knowledge to the public that choses to use them in spite of the risks.

Drug companies are being punished for developing novel first-in-class new drugs and criticized when they do not. They spend enormous amounts of money to assure that their products are safe and effective and undergo what can only be considered arduous scrutiny by the FDA to earn the right to sell their products. When they misrepresent their products by withholding data from the agency they should to be punished to the full extent of the law. The majority of drug litigation cases do not constitute such behavior. On the contrary, most are examples of pure opportunistic greed only directed at the sponsors and not the FDA that is totally complicit. Yet, in most cases there are two big losers: the company and the patient. This needs to change if we want to accelerate the progress toward cures. There has to be a degree of trust in the drug companies that will permit them to operate on an innocent until proven guilty basis and not the reverse as is now the case. The ambulance chaser mentality that currently exists has forced the entire medical R&D establishment into a defensive mode.

MALPRACTICE MADNESS; WHO WINS?

There was a time when physicians were a trusted and admired group. People knew their doctor like a family member and trusted in their judgment. They practiced their art with a keen sense of compassion and took the time to actually come to the patient. They shared the risk–benefit decisions of disease with the patient and were not focused on the legal consequences of things turning out badly. My Mother died of breast cancer at the age of 29 because a physician misdiagnosed a lump on her breast. No one thought of suing the family physician who was using his best judgment that turned out to be wrong. Today the story would be much different, a fact that has given rise to defensive medicine and five figure medical malpractice insurance premiums for physicians. This relationship and trust between the patient and physician is critical to the practice of medicine, and it has become a thing of the past.

Surgery is not nearly as heavily regulated as drug therapy and yet it is much riskier than drug therapy. Surgeons are relatively free from the system that blocks other medical subspecialties, and the reason is that they are generally less dependent on drugs for the treatment of their patients. They are also relatively free from the demoralizing subservience to managed care. Even the pain medication associated with surgery is administered by anesthesiologists, not the surgeon. The surgeon is a relatively free agent, as all physicians should be. The main reason for the startling advances in surgical techniques is that they are relatively free from both regulation and liability and not slaves of defensive medicine.

Drug litigation and medical malpractice lawsuits have made physicians timid and introduced the concept of defensive medicine and that is not what you want when you are starring death in the face with a terminal disease. Malpractice insurance rates have skyrocketed solely due to the tort system. Physicians opt for conservatism in the face of risk. This is fine when dealing with routine chronic disease therapy but it is literally deadly when the issue is terminal disease. Many physicians avoid high-risk patients for fear of having to make life and death decisions that could fail. This shifts the medical profession toward patients that need it the least and away from those who need it most. Anything that makes physicians shy away from treating terminal disease patients aggressively will surely delay progress toward curing them.

THE RISK–BENEFIT CONUNDRUM

If a drug is found to increase the risk of heart attacks by 2%, there is no way of knowing which heart attack patients actually had a drug-induced heart attack, yet the company is liable for all the heart attacks, not just those caused by the drug. There is just no way to establish cause and effect. This rationale was used to the advantage of the tobacco companies with lung cancer, but it works against the drug companies. They are liable for all the heart attacks if data are available showing an increased risk associated with their drug. This is an enormous price to pay and drug companies plan their business strategy and drug pricing based on it.

If a drug causes 50 additional heart attacks in 20 million patients but saves 25,000 lives per year, is that something to sue over? The risk to 50 is the benefit to the 24,950. How should it be dealt with? Might those patients who die from what the drug treats sue those that forced the company to withdraw the drug from the market or the company itself? What are the rights of the patients whose lives are saved by drugs that are prematurely pulled from the market because of class action lawsuits? The drug companies get caught in the middle of these ethical issues that they and the FDA must wrestle with on a daily basis. Unfortunately these battles are fought in the emotionally charged atmosphere of a litigation that involves a patient caught on the wrong side of any drug's spectrum of effects. Those that benefit from the drug are rarely heard. This is a very complex issue with no simple answers and certainly the perfect situation is one where no one is harmed by a drug. If this were required of all drugs there would be no drugs. Certainly me-too drugs should be scrutinized very closely for fatal side effects, and those with the greatest occurrence of such effects should be withdrawn or carry a black box warning on the label but liability is not the best solution to this problem.

LIABILITY FOR FAILED CLINICAL TRIALS

An additional problem that all public pharmaceutical and biotech companies must endure is the scenario of a failed clinical trial that causes their stock price to decrease dramatically. This often results in class action lawsuits, which is really absurd. Clinical trials are extremely risky and liability for a failed trial puts an enormous burden on companies already punished by a large decrease in their market value. Clinical trial failure is especially hard on small companies that generally have only one late-stage product. Defending against class action lawsuits often renders them unable to try the drug in another disease or modify the failed trial.

Until a better way of dealing medical liability is found it will be a barrier to the entire practice of medicine and will hinder the conquest of terminal disease the most.

REFERENCES

[1] R.S. Bresalier, R.S. Sandler, H. Quan, J.A. Bolognese, B. Oxenius, K. Horgan, C. Lines, R. Riddell, D. Morton, A. Lanas, M.A. Konstam, J.A. Baron, Cardiovascular events associated with rofecoxib in a colorectal adenoma chemoprevention trial, New England Journal of Medicine 352 (March 17, 2005) 1092–1102. http://www.nejm.org/doi/full/10.1056/NEJMoa050405.
[2] A. Berenson, Merck agrees to settle Vioxx suits for $4.85 billion, New York Times (November 9, 2007). Retrieved from: http://www.nytimes.com/2007/11/09/business/09merck.html.
[3] Vindicating Vioxx(Editorial), The Wall Street Journal (May 31, 2008). Retrieved from: www.wsj.com/articles/SB121218857343634215.

[4] H.A. Waxman, The lessons of Vioxx – drug safety and sales, New England Journal of Medicine 352 (June 23, 2005) 2576–2578. http://www.nejm.org/doi/full/10.1056/NEJMp058136#t=article.

[5] S.E. Nissen, K. Wolski, Effect of rosiglitazone on the risk of myocardial infarction and death from cardiovascular causes, New England Journal of Medicine 356 (June 14, 2007) 2457–2471. http://www.nejm.org/doi/10.1056/NEJMoa072761.

[6] J. Goldstein, An Avandia lawsuit emerges, inevitably, The Wall Street Journal (June 12, 2007). Retrieved from: http://blogs.wsj.com/health/2007/06/12/avandia-lawsuits-emerge-inevitably/.

[7] B. Richter, E. Bandeira-Echtler, K. Bergerhoff, C. Clar, S.H. Ebrahim, Rosiglitazone for type 2 diabetes mellitus, Cochrane Database System Review (3) (July 18, 2007). Retrieved from: https://www.ncbi.nlm.nih.gov/pubmed/17636824.

[8] G.A. Diamond, L. Bax, S. Kaul, Uncertain effects of rosiglitazone on the risk for myocardial infarction and cardiovascular death, Annals of Internal Medicine, Ideas and Opinions 147 (8) (October 16, 2007) 578–581.

[9] A. Edney, Glaxo's Avandia cleared from sales restrictions by FDA, Bloomberg (November 25, 2013). Retrieved from: http://www.bloomberg.com/news/articles/2013-11-25/glaxo-s-avandia-cleared-from-sales-restrictions-by-fda.

[10] R. Weisman, Biogen Idec, Elan facing suits over MS drug side effects, The Boston Globe (September 13, 2013). Retrieved from: www.bostonglobe.com.

[11] H.I. Miller, The saga of silicone breast implants, Forbes Magazine (March 4, 2015). Retrieved from: http://www.forbes.com/sites/henrymiller/2015/03/04/infuriating-titbits-about-silicone-breast-implants/#53c37fe18d60.

[12] G. Kolata, Panel confirms no major illness tied to implants, New York Times (June 21, 1999). An independent panel of 13 scientists convened by the Institute of Medicine at the request of Congress has concluded that silicone breast implants do not cause any major diseases. Retrieved from: http://www.nytimes.com/1999/06/21/us/panel-confirms-no-major-illness-tied-to-implants.html.

[13] Risner, et al., Efficacy of rosiglitazone in a genetically defined population with mild-to-moderate Alzheimer's disease, The Pharmacogenomics Journal 6 (2006) 246–254. Retrieved from: http://www.nature.com/tpj/journal/v6/n4/abs/6500369a.html.

Part B

Towards Solutions

Chapter 7

Cleaning Up the Academic Research Quagmire

Chapter Outline

Academia is getting away with one of the biggest charades in modern times. They have managed to convince the rest of us that their existence is immune to the checks and balances that dictate all the other activities in our society. We have been coerced to revere academia and allow it to dictate that parents and students pay their prices for an education that is becoming increasingly less valuable. Even public universities are becoming beyond reach pricewise, and student loan debt now exceeds a trillion dollars. They play the same game on the research side, and it is time that they become accountable and answer to the same metrics of success and results that apply to all the other sectors of society. The price paid for this privileged status when it comes to curing terminal disease is measured in lives, and it is time to demand accountability.

Finding cures for the major terminal diseases is too important to trust to the relaxed, unorganized, and unaccountable world of academia. There must be a national strategy that establishes a timeline to achieve preestablished goals. Scientists are immune to answering the needs of the marketplace or society. We see this mentality across the entire research spectrum, including universities, institutes, and foundations, and it is killing us. Medical science must emerge from its cloistered existence and join the mainstream of society. We trust it to thousands of scientists with other agendas and an array of institutions totally untrained in how to solve complex problems. It is amazing that this has been tolerated for all these decades.

A Roadmap for Curing Cancer, Alzheimer's, and Cardiovascular Disease.
http://dx.doi.org/10.1016/B978-0-12-812796-4.00007-6

REBUILDING THE NATIONAL INSTITUTES OF HEALTH: NASA FOR CURES

The National Institutes of Health (NIH) was conceived as the center where medical science would be translated into better health and I spent over 12 years of my career there. Obviously medical research should produce cures first, but the universities where much of this research is typically done focus on other things such as academic freedom, science for its own sake, teaching, tenure, and above all publishing. This dilutes the search for cures so the thinking was that the government should create a center where researchers would be free of these academic duties to focus their efforts on health research. In the NIH mission statement there is no mention of cures (*Mission and Goals from the NIH website. NIH's mission* is to seek fundamental knowledge about the nature and behavior of living systems and the application of that knowledge to enhance health, lengthen life, and reduce illness and disability). The mission talks about loftily and nebulous academic goals rather than the specific goal of curing the diseases that are sure to kill. The mission implies an unfocused academic wandering through basic life science rather than a priority-based purpose to rid society of the most deadly of diseases. This is exactly what has played out over the past 60 years.

After 60 plus years there has been precious little bottom line results regarding the terminal diseases other than millions of scientific papers, the great majority of which have nothing to do with disease. The NIH was the right idea but the wrong execution since it is basically governed and staffed by academics essentially allowing scientists to pursue their own professional goals rather than serve a larger more urgent mission. It is a public university without students. Certainly removing the teaching from research was a step in the right direction but only a marginal one. The far more important issue of infusing urgency, accountability, and strategy into curing disease is not even alluded to. The results speak for themselves; there has been a lot of science but virtually no cures. This is in spite of a budget that has increased rapidly, especially between the years of 2000 and 2003 when the budget nearly doubled. One might ask what effect that has had on cures, but the answer is disappointing. Doubling spending 14 years ago and the war on cancer 47 years ago has had little effect on cures; this in itself should be a wake-up call that something needs to change at the NIH.

Each of the 27 NIH institutes was formed at a different time. Interestingly, the cancer institute was formed first in 1937 followed by the Heart and Lung Institute in 1948 and the Neurology Institute in 1950. The idea seemed right at the time and they should have stopped there. So the three institutes that were supposedly dealing with the diseases that to this day over 60 years later are killing 2 million people each year are still here. If the begetting of new institutes stopped in 1950 and the focus, strategy, and management were appropriate, we would be in a very different place today, but politics took over and every interest group had to be served, which resulted in the enormous bureaucracy we have today. NIH must be transformed from an academic unfocused 27-institute politically driven monument to science into a laser-focused project-oriented entity with the sole objective to cure the three diseases that are killing the most people, all of which dwarf the health issues dealt with by the other 24 institutes combined.

This new and streamlined *terminal disease administration or TDA* would consist of four institutes with an entirely different mission statement: "The TDA will cure the terminal diseases for which there are currently no effective treatments or cures." There would be one institute for each of the major terminal diseases; cancer, degenerative, and cardiovascular. A fourth institute would consist of a regulatory component for each terminal disease separate from the existing Food and Drug Administration (FDA), which could now focus solely on drugs for the lower priority currently treatable diseases. These TDA regulators would monitor safety and efficacy concerns on site with a totally different set of metrics designed to serve terminal disease patients and their unique risk–benefit parameters. The TDA would answer directly to a newly created cabinet head separate from the department of health and human services.

The drug companies would deal directly with TDA regulatory staff with their terminal disease drugs. This key reform would transform the regulatory mission strategy to one of cooperation rather than antagonism with the regulatory authorities and add a laser focus on terminal disease rather than the old NIH focus on health research in general. It is critically important that scientists employed at the TDA are crystal clear on what their goal has to be and not permitted to wander through basic science pursuing their academic freedom as they are now permitted to do. They will have to become practitioners who apply science as a tool toward an end and not be obsessed with publications, tenure, and academic freedom, luxuries that only cloud and greatly slow down progress.

The TDA will also replace the current academic scientist management with project-focused goal-oriented managers to institute a strategic milestone-driven operational culture that would be accountable to specific drug development progress metrics. There would be enhanced focus on applied research focusing on innovative uses of existing science and drugs as a near-term strategy for generating potential cures or life-extending therapies. New avenues of converting cancer and Alzheimer's disease into manageable conditions would be exploited employing many more options involving diet, using old drugs, and combinations thereof in new ways as well as physical interventions such as ultrasound and various forms of radiation. Terminal diseases will be attacked from every angle; more stones unturned must be the goal not just beating one science theory to death.

New science will continue to be vetted as a longer term strategy but put in a more rational context as being high risk and unproven rather than viewed as the cure we all have to wait decades to see. The basic research component is important to continue, but under the more rigorous TDA management it will be accountable to generally accepted milestone guidelines and conducted in an organized fashion rather than subject to the whims of academic freedom. The same will be true for all extramural research funded by the TDA. Grantees will have to adhere to strict guidelines for the work they want funded and demonstrate a clear focus on the disease and adherence to mandated quality control procedures. Grantees at universities and private institutes receiving extramural TDA funding must insure that the proposed work is not redundant with work from other institutions and is strictly focused on the disease.

All work at the TDA, both intramural and extramural, would now have to be quality controlled using the good laboratory practice (GLP) and good management practice. This is absolutely critical to insure that laboratory data is real and able to be reproduced, an increasingly major problem [1,2]. Quality control requirements will also reduce false starts on the development side and the high failure rate in start-up companies that often license such work. This quality control requirement for laboratory research will add cost to research in general and some if not all of that cost should come from the overhead/indirect cost each university charges the NIH, which currently is obscene [3–6]. The TDA will set a fixed cost for grantee universities and private institutes when it comes to indirect costs. This will be in the neighborhood of 15%–20% and not the current 50%–90% range that currently exists. This will force academia to control expenditures and enable more direct support to research.

The TDA would also have drug synthesis capabilities and other support functions involved in drug development such as formulation development, toxicology testing, and stability analysis. This would transform the current NIH into a much more relevant engine for identifying actual drug candidates for preclinical testing and early-stage clinical testing. The clinical center facilities at the TDA would also be expanded using the current facilities currently occupied by the other 24 institutes. The clinical trial capabilities will be reformed to meet industry standards; this is absolutely critical. Clinical trials conducted at the current NIH have been problematic in a manner that is consistent with the scientists who manage the institution [7–10]. A wide range of quality control issues have plagued the NIH clinical center that must be corrected before additional trials are conducted. The kinds of deficiencies found at the NIH clinical center would have shut down a drug company, yet the media has been kind to the NIH because of its universally prestigious designation that it shares with academia in general. The media needs to get in line here, so the public sees these serious issues. It has long been known in the industry that when one agrees to do a clinical trial at the NIH that it will never get done on time and it will become a science project that the FDA will probably not accept as a pivotal efficacy trial. That is why companies will rarely sponsor NIH clinical trials. This will change at the TDA.

The extramural research budget of the current NIH, which provides research grants to academia, would be revised to not only insure quality control issues but also provide funding for patent filings. A key goal of everything that goes on in both intramural and extramural work will be the filing and prosecution of patents and not publishing papers in the literature. As already discussed, it is useless to publish before patents are filed in the real world of drug development. This will make both intramural and extramural work more attractive for industry licensing and more valuable to entrepreneurs in general for commercialization.

A critical added feature of the TDA would be its mandate for greatly enhanced partnering with industry in ways that would facilitate the transfer of high-quality data into products. Rather than having venture capitalists (VCs) fund the licensing of technologies out of the NIH into start-up companies and build redundant infrastructure at these new companies there would now be a more efficient and rapid transfer of technology into development. Having industry directly involved with TDA research, both intra- and extramural, rather than go through the middleman of VC funded start-ups will obviate the need for technology transfer of research into products, it will be seamless and automatic. The drug companies are pros at developing products, while start-up companies are students who hire retired industry pros anyway. During my years at the NIH it became increasingly more difficult to engage in outside activities and collaborate with industry, something that was very frustrating to me. I find from speaking to old friends still there that industry collaboration has now gotten even harder. This is counterproductive to technology transfer, a process that is now in vogue but is clumsy and slow.

The TDA will be a much more efficient engine of new drug discovery and development that will be the equivalent for cures to that of NASA for space travel. It will bring all the components of biomedical research under one strategic umbrella and be the epicenter of what is now a totally disorganized medical R&D universe fraught with inefficiencies, redundancies, and arcane inappropriate regulation. The current bloated inefficient clashing of individually flawed bureaucracies was never designed to deal with the cure crisis but rather with a much more nebulous mission of advancing health science in a haphazard indirect manner. Injecting more capital into the current NIH has already been shown to be futile and is not the answer. In 1998 the budget for the cancer institute was tripled, again there is little to show for it 19 years later. Now the 21st Century Cures Act is proposing immediate increases to the NIH budget of over 15%. It joins a long list of futile efforts since money is not the holdup with cures, the system is.

The proposal put forth here regarding reform of the NIH and transforming it into the TDA should be considered as a first draft and certainly the details need to be worked out and put into practice. The history of the current NIH is clear; it has not delivered what is needed in spite of its dramatic growth and almost a century of existence. It may succeed eventually without reform but to accelerate the process and see some major progress for the current generation we need to get serious about this institution. My years at the NIH are fondly remembered as is my love for what it represents, so none of this is meant to demean it but rather to focus it on the acute needs of humanity. Some of the hardest working and talented scientists are at the NIH, and their publishing productivity makes it one of the most productive government agencies. The problem is that it must transition to a cure-focused organization with all the tools and leadership required to find answers.

WHAT TO DO WITH ACADEMIA AND INDIRECT COSTS?

It is not just the NIH that needs to be reformed and infused with strategy, goal-oriented accountability, and leadership. There are thousands of universities, foundations, and institutes throughout the country and the world that are engaged in medical research that is similarly plagued with the same problems. The metrics by which these centers are evaluated are nonexistent. Few, if any, have made major breakthroughs that have changed treatments. None have cured a major terminal disease, and yet virtually all continue to grow, prosper, and become "prestigious" not for what they have done but more for who they are and how many awards their staff have received. Academic freedom at universities will prevent any management initiatives that direct the activities of their professors, and this is probably a good thing for academia that claims to support such a mindset. That freedom, however, leads to chaos when a defined goal needs to be achieved. A case can be made that universities and academia in general should be taken entirely out of the quest for cures given all the diversions that academia presents. Granting agencies and the capital they control can be put to much more efficient use when deployed to organizations that are purely focused on a specific goal and do not have to compromise their primary mission.

Academia is a very expensive venue for research. The revenue streams of universities are enormous and need to be seriously audited and made available to the students and parents paying tuition bills. Just about all of the scientific staff at universities is entirely supported from external grant funding. Grants always pay the salaries and benefits of the grantees. Those who are not getting their salaries fully funded from grants will not get tenure. Grant writing is a never ending job for scientists, and finding a cure for cancer would put many of them out of business. This coupled with the huge overheads or indirect costs that universities, hospitals, and not-for-profit demand from granting agencies raises serious questions regarding the efficiency of how research money is deployed. If one then factors in the huge endowment funds that most universities accumulate from alumni and other sources, one begins to wonder whether there is some kind of racket in progress. The average overhead charged to granting agencies such as the NIH by universities and private institutes is about 50% and can be as high as 100% [3–6].

This entire fiasco must be reined in because many billions of dollars of research money are being wasted. The grantee should be paid indirect costs, but this apparent usury on the part of academia must be stopped. A partial solution would be to have the grantee institutions provide a quality control function for all medical research and mandate that all such work be strictly disease focused to qualify for federal grants. Other granting sources must adopt similar caps on indirect cost amounts rather than just pay what universities dictate. So universities do not have to pay their science faculty and skim enormous amounts of money off the top of research grants; you may want to consider this the next time the call comes from the alumni society of your Alma Mata.

CHARITIES: YOU HAVE TO BE KIDDING!

The folly of the entire cure enterprise is best exemplified in the numerous charities that exist to find cures. Imagine if we funded the Moon launch with charities. Not one charity or cancer walk or telethon has succeeded in finding a cure and it is time we all look at them with a microscope before writing any more checks. They simply add to the confusion and exist to make people feel good about helping the cause. Possibly the donors are sublimating their own angst about the suffering of the stricken individuals, but pouring money into charities has not been productive. Some provide support for those afflicted with a given disease and that is important but those who offer the hope of cures have not produced. It insults those actively involved in medical research to see that charities are necessary to do what they should be doing and maybe that is a good thing. The larger message it sends to the public is that we are doing something wrong if charities are needed.

The Muscular Dystrophy Association (MDA) Telethon is the best example of waste in medical research; what has decades of those extravaganzas produced? Are we all so naive to still believe that charities will produce results in such an urgent and complex task? It is time to put our money into the right efforts and charities are not it. Certainly the Red Cross, St Jude's Hospital, and related organizations as well as the Salvation Army are worthy of support since they directly help people. Terminal disease does not fit for charities and is in fact an insult to the cause.

The same logic applies to donating to your Alma Mata, probably the biggest charity of all. The nerve of universities to ask alumni for money every year is remarkable. They have priced a university education out of reach of most families to the extent that most graduates are faced with college loan debt that takes years to pay off. Parents paying the tuition are either planning a very short retirement or second mortgages on their homes. What are universities doing to justify those contributions? They certainly are not curing anything and more likely simply exploiting the donor's philanthropic sense under false pretenses. They are cloning money and certainly do not need any more.

QUALITY CONTROL IN RESEARCH

Researchers labor long and hard to generate their data. When they get it, they rush to publish it for fear of being scooped. They chose the statistics they want to use. They see the results and are in most cases free agents regarding how they want to interpret and reduce it to conclusions. They have no training in quality control, and no one oversees what they do. This creates a host of problems, first and foremost of which is the poor reproducibility of basic science work. The rush to publish an exciting finding is very difficult to overcome when one's career is dependent on it. The lure of peer acclaim is also a powerful thing, and I have seen it played out in blatant fabrication of data to get something published. While this is a rare event, the pressure to publish often results in poor quality control that perpetuates premature and difficult-to-reproduce data. Whatever the cause, the problem of poor quality control in medical research is a big issue that needs correcting if one is to take the data seriously [1,2].

The journals are full of premature findings that go unreplicated by other laboratories. There is in fact a journal dedicated to this called "The Journal of Irreproducible results." One is therefore routinely confronted with having to choose which laboratory to believe, if any. Erroneous findings may persist for decades until it is determined which laboratory is wrong. We have all seen this over the years. One day caffeine is great and the next it is not. Stem cells are potential cure-alls, yet no one talks about their propensity to produce tumors, the list is long. This poor attention to data quality garbles the entire medical research landscape and results in numerous false starts when venture capital start-up companies are formed to pursue these findings. Unfortunately, the journal editorial boards that decide which research papers will be published do not dwell sufficiently on issues relating to quality control. Scientists view themselves as deep thinkers and the notion that they should engage in the bean counting activity of quality control is both insulting and foreign to them. They focus on the science and how novel it is and not how impartially the author is interpreting the findings or how many times the studies were repeated and whether the investigator was sufficiently blinded regarding which data points were controls. Details regarding methods and experimental design are often overlooked. Too few scientists attempt to reproduce findings from other laboratories since Nobel Prizes are not awarded for that.

The remedy for the quality control problem in academic research is not going to be popular or easy to achieve. The medical and biochemical journals need to tighten up their requirements and require all published papers to be conducted under GLP-like conditions. Grants must insist on this as well and provide the funding for a QC capability with every grant. This would result in a lot less published papers and much shorter curriculum vitas but would greatly improve the quality of the basic science literature. Journal publishers and academics would both initially suffer. It is not uncommon for a thought leader or a productive researcher to have hundreds of published papers in their CV. While this does wonders for the ego, one wonders what it all means when in most cases so little tangible results that can solve disease-related problems is contained in these voluminous bibliographies. Instead of providing such lucrative indirect cost amounts to support university infrastructures, research grants should provide funding for data management and quality control to investigators. Fees for publishing papers would have to be increased to provide for the data management and quality control specialists required on editorial boards. Research institutions and universities that receive grant funds would have to provide support staff for quality control and management specialists to assist researchers in improving the quality of their data and experimental design.

INJECTING STRATEGY AND MANAGEMENT INTO MEDICAL RESEARCH

There would be no shortage of grant funds for universities and medical research institutes if they are properly directed, not used to fund haphazard research and if the grantors got tough on the universities and institutes on the indirect cost issue. Just putting a cap on indirect costs of 20% would greatly increase the number of grants awarded. The lack of coordination, strategy, and management regarding the way research is currently funded is mind boggling. The best analogy is again the Keystone Cops; lots of uncoordinated activity but not much to show for it. Making researchers and academic institutions accountable and improving the quality and reproducibility of their work is important, but it will make little difference if the system of funding and conducting medical research is not coordinated and managed in a goal-directed manner. One granting agency does not know what the other is doing; they compete for attention rather than plan together for a shared goal. While this works in many areas of commerce, it is a disaster when complex problem solving is the goal and decades of effort have been unsuccessful. The resources are not the problem; this is what makes the situation so ridiculously frustrating, no one wants to talk about the real issues but rather just pump more money into a chaotic mismanaged collection of vested interests.

How do we proceed to reinvent the medical research system to facilitate cures? Clearly, in its present form it is doomed to slow progress. The first task is to segregate long- and short-term opportunities for terminal disease cures and establish clear goals in each area. The long-range cures will most likely emerge from basic research, i.e., stem cell and gene therapy, to name just a few. It will take a long time for the basic research to produce results, but the shorter term applied research will produce quicker results or at least reduce disease progression or hold it at bay as was the case with AIDS. The short-term cure opportunities will be more innovation driven and involve using existing science in novel ways such as old drug–new use scenarios or trying new approaches to disease mitigation as was done with peptic ulcers. This applied research has the potential of saving lives in the near term and this is what is most needed. Resources and talent must be rationally allocated to each of these research areas in accordance with a well-crafted strategy that avoids duplication and greatly increases the number of distinct approaches that can be entertained.

Most basic research occurs at institutes and universities, so they must clean up their operating mentality before being entrusted with such an important task. Much has been written about what has happened to universities since the 1960s [11–13] and that requires a separate book but lets focus on the issues most important to cures. First, they must reduce and earn their indirect costs and use the funding they get from grants to provide more useful support to their staff in the areas of quality control, intellectual property, preventing redundancy, and experimental data analysis and design. If a university wants to receive grant funds in a particular area such as cancer research, it must have a dedicated support staff with drug development expertise in that area. They must communicate with their counterparts at other universities on a regular basis regarding other ongoing work in that field. Metrics to evaluate progress, prevent redundancy, and foster collaboration must be adopted. Part of the indirect costs from grants would support this function. This group of cancer research coordinators would report to the major granting sources twice a year to discuss progress and plan basic research strategy with a focus on efficiency, productivity, accountability, and determining what is working and what should be shelved. The same strategy would apply to the other two major terminal disease categories, cardiovascular, and degenerative disease. The granting agencies should also dedicate a portion of their funds to support these efforts. The clear purpose of these reforms is to get our enormous engine of research talent and infrastructure under one hood and pulling one train rather than perpetuate this haphazard inefficient research landscape characterized by academic fiat and hoping that something relevant might fall out of the process.

Research institutes may be a little easier to reform than universities given their private character and the absence of a major teaching mission, but the same things need to happen as was discussed for the NIH and the universities. What needs to happen is similar to what we did in the space program and the Manhattan Project where scientists were coordinated, focused, and asked to sacrifice their academic freedom and achieve a mission that transcends their academic aspirations. The TDA should be the center of the newly reformed research establishment, the hub of the wheel that navigates the effort to success. It is time to give this quest for cures the serious and urgent character that has been so successful in building modern civilization and stop relying on academia, charities, and false hopes.

FEWER THOUGHT LEADERS AND MORE MAVERICKS

The contribution of mavericks to curing disease has been impressive. Mavericks see the world through a different lens. The thought process they practice departs from conventional wisdom, and this is particularly important in the medical sciences. Chemistry and physics are the most quantitative of sciences. Biology is different and far less precise with laws that are highly variable by comparison to the exact sciences. Chemistry and physics are not dictated by a genome that is constantly being reshuffled. Qualitative judgments and innovation are more relevant in biology due to the enormous variation inherent in biological systems. This is what led me to biochemistry, a dynamic landscape that is full of surprises. A lifesaving drug for one patient can kill another. A drug that works 1 day may not work the next. Medical research data is always harder to interpret and can often mislead the researcher to false hypothesis generation. Thinking outside the box takes on a whole new meaning in biology because the boundaries of the box are much less clear than in other sciences.

Thought leaders are the politicians of science, and they generally get their start by generating a preliminary finding and marketing it in a manner that generates a buzz in the academic community. They are always careful to say that the findings are preliminary, but the media, whose job is to keep us entertained, swoops in and the frenzy of waste and futile hopes begins. These preliminary findings often become the research areas that research foundations and institutes get built around. Gene therapy, biotechnology, stem cell research, mapping the human genome, and cloning are the exciting findings that perpetuate thought leaders. This leads to enhanced funding of thought leader projects at the expense of other less visible investigators. The bottom-line result is a very unhealthy focus of research in relatively few areas and a great deal of expensive dead ends. This produces megaliths in the form of institutes that become self-perpetuating. Precisely the opposite is what is needed, especially in diseases we have failed to cure after decades of trying. A much greater scope of efforts is necessary. Lots of work on a few hyped ideas is not the desired path. More stones need to be turned over when success has been so elusive, and thought leaders inhibit that dynamic while mavericks thrive on it.

Thought leader–driven science is also counterintuitive to creativity in general since it runs counter to new hypothesis generation. More effort is directed to fewer hypotheses, and idea boxes are built in which researchers become trapped in. It is a form of regimentation that dictates funding priorities and inhibits new ways of looking at old problems. This has been especially true for Alzheimer's research [14–16]. So, what do we do with all the thought leaders who dictate the direction of mainstream medical research? One answer is to encourage more maverick science by increasing its support. This is easily accomplished if more virgin hypotheses get funded. As we have seen. Mavericks often have to swim upstream to get their ideas heard. The Australian researcher who had to self-induce peptic ulcers to demonstrate his hypothesized cure to his peers is sad proof of this dilemma.

The fuel of thought leaders are science-based awards, peer worship, media attention, and tenure; all of which is counterproductive and need to be abolished.

They are self-serving and neither has ever been shown to encourage innovative science or to be correlated to curing a disease. In most of the cases I have seen, tenure leads to a coasting mentality where the professor becomes comfortable with and protected by their past and confident that the future resides in a pension and the lecture circuit rather than accomplishments going forward. Nobel Laureates consult for VCs and join prestigious boards where they exert a pervasive influence on the directions that both small and large companies pursue in their research and development efforts. Popular science has not been productive, and we need to jump-start new ideas and approaches and not feather the nests of those that now control the direction of science and drug development. Awards and tenure provide the wrong incentives and play a major role in creating what has been the greatest stranglehold on innovative drug development, the thought leader. Thoughts are not to be led, especially those we are in the most need of right now, innovative thoughts.

THE PUBLIC RIP OFF

The public has been taken for an expensive and tragic ride that plays out in virtually every family, terminal disease. Citizens have given generously to support medical research and received nothing in return other than being a lifetime target on a mailing list and recruited to take long walks for a cure. We give in numerous ways: foundations, telethons, cancer walks, our alma maters, and our tax dollars to name a few. We do not understand where it is all going and what it supports but clearly, the facts show a horrible return on investment. We still live in fear with little on the horizon other than the much touted science breakthroughs that the media thrives on. At the very least we must demand that this money we give be used in a much more efficient manner. The Muscular Dystrophy Telethon is one example; others are the numerous cancer societies, the NIH, the impressive research institutes with their thought leader academics and many others. We keep giving and get very little other than the false hope that the huge enterprise of academia and science is on the verge of curing the diseases we all dread. It is not happening and the public can change that. There must be a public outcry to make this a national issue with a priority equal to that of providing for the national defense.

Medical research must be transformed if we are to speed the discovery of cures. All these "prestigious institutions" are not producing; their biggest success is raising money and elaborate architecture. That designation comes from the thought leaders who reside in them and the exciting science they peddle that never produces results. Prestige must emanate from real-time performance and that is nowhere to be found in cures research. It will require creative and fundamental actions to fix this flawed medical research establishment and finally end this greatest of human tragedies.

REFERENCES

[1] G. Naik, Mistakes in scientific studies surge, The Wall Street Journal (August 10, 2011) A1–A12.

[2] A. Marcus, Lab mistakes hobble cancer studies by scientists slow to take remedies, The Wall Street Journal (April 21, 2012) A1–A12.

[3] F. Mussano, R.V. Iosue, Colleges need a business productivity audit, The Wall Street Journal (December 28, 2014). Retrieved from: www.wsj.com/articles/frank-mussano-and-robert-v-iosue-colleges-need-a-business-productivity-audit-1419810853.

[4] A. Cave, Taking a hard look at University research, Stanford Social Innovation Review (October 20, 2014). Retrieved from: https://ssir.org/articles/entry/taking_a_hard_look_at_university_research.

[5] H. Ledford, Indirect costs: keeping the lights on, Nature (November 9, 2014). Retrieved from: http://www.nature.com/news/indirect-costs-keeping-the-lights-on-1.16376.

[6] V. Callier, Overspending on overhead, The Scientist (February 1, 2015).

[7] R.B. Howard, I. Sayeed, D. Stein, Suboptimal dosing parameters as possible factors in the negative phase III clinical trials of progesterone in TBI, Journal of Neurotrauma (2015) http://dx.doi.org/10.1089/neu.2015.4179. (E pub ahead of print).

[8] NIH Clinical Center Riddled with Research Problems, AHC Media, August 1, 2016. Retrieved from: https://www.ahcmedia.com/articles/138301-report-nih-clinical-center-riddled-with-research-problems.

[9] T.M. Burton, Mysterious fungus sparks crisis at NIH, The Wall Street Journal (January 19, 2017). Retrieved from: http://www.wsj.com/articles/mystery-fungus-sparks-nih-crisis-imperiling-trials-patients-and-its-boss-1484753489.

[10] Draft Report, The Clinical Center Working Group Report to the Advisory Committee to the Director, NIH. Reducing Risk and Promoting Patient Safety for NIH Intramural Clinical Research, April 2016. Retrieved from: http://bit.ly/29Qjm9a.

[11] N.S. Riley, Reading, writing, radical change (review for the book crisis on campus, by M. C. Taylor), The Wall Street Journal (August 31, 2010) A15.

[12] M. Bauerlein, Ignorance by degrees (review for the book higher education by A. Hacker & C. Dreifus), The Wall Street Journal (August 2, 2010) A11.

[13] J. Piereson, N.S. Riley, The Ivy League doesn't need taxpayers' help, The Wall Street Journal (October 17, 2016). Retrieved from: www.wsj.com/articles/the-ivy-league-doesnt-need-taxpayers-help-1476652968.

[14] S. Begley, Is Alzheimer's field blocking research into other causes? The Wall Street Journal (April 9, 2004). Retrieved from: www.wsj.com/articles/SB108145279348578177.

[15] S. Begley, Scientists' world-wide battle a narrow view of Alzheimer's cause, The Wall Street Journal (April 16, 2004). Retrieved from: www.wsj.com/articles/SB108206188684384119.

[16] E. Lilly, Company, Leading Experts Warn Global Goal to Cure or Treat Alzheimer's Disease by 2025 Is at Risk, Offers Solutions to Adjust Course, PR Newswire, October 11, 2016. Retrieved from: www.prnewswire.com.

Chapter 8

Reforming the FDA

Chapter Outline

The Food and Drug Administration (FDA) has total control over the drug industry. Who would want to start a business with that hanging over their heads: A business where only a small percentage of products teed up make it to the market; a business where the cost of developing that product is subject to a regulatory agency that has no concern about the impediments it imposes on the development path of your product; and a business that is prone to not being trusted by this agency that controls its destiny. Well this is the position virtually every drug and medical device company finds itself in. The FDA is certainly justified to protect the public, but when they become an obstruction to the quest for cures to diseases that kill with certainty, something has to change.

The FDA started out as just a regulator of drug safety and left it to drug companies and physicians to determine whether it worked in the real world but in the 1960s its mandate was extended to include efficacy. This change was largely because of an overreaction by Congress to the Thalidomide problem discussed earlier. The inclusion of efficacy was warranted in the case of drug products anyway and the quality of such has improved because of it, but it has become increasingly obvious that the agency has reached a point where regulation has evolved to strangulation. This is especially true in the realm of terminal disease drugs as was discussed in Chapter 5. Examples of the FDA holding up the approval of important drugs, and the effect this has had on both the cost of drugs and human lives are already common knowledge [1–7].

THE EVOLUTION OF DRUG REGULATION

Government regulation of drug safety and efficacy is a good thing, but it has gone way too far for all drugs, especially when it comes to terminal disease.

A Roadmap for Curing Cancer, Alzheimer's, and Cardiovascular Disease.
http://dx.doi.org/10.1016/B978-0-12-812796-4.00008-8

Academism, an inordinate distrust of industry and the total usurping of a patients right to life have virtually put a choke hold on drug approvals. Patients and their physicians often are at odds with the FDA when it comes to accessing potentially lifesaving investigational drugs, a situation that has resulted in the formation of advocacy groups such as the Abigail Alliance [8], the right to try movement [9], and others that seek to gain access to experimental new drugs for hopelessly terminal cancer patients. The "right-to-try" movement that is currently sweeping the country is a clear expression of the frustration with overreaching drug regulation. While 28 states have now passed right to try legislation, the FDA has yet to address whether it will comply and even if it does it will set the parameters, which likely will be stringent. The US government also passed a right to try bill [9]. The right-to-try bills are all State sponsored currently and federal laws are being proposed, and the FDA will probably override them requiring more stipulations for patient access but at this point they are a welcome expression of the frustration that is building. These bills advocate availability of an investigational drug after Phase I trials [10] as some have advocated before [11,12], and this should be changed to after Phase II when some degree of efficacy is established. The drug companies will likely be more willing and able to engage in that, and the liability issues will be less. The FDA should be facilitating lifesaving drugs, but yet it brazenly maintains a strangle hold on their availability while the public is relegated to being treated like children unable to make the most fundamental of decisions. It remains to be seen how these right-to-try State and federal bills will play out.

The Hippocratic Oath preaches that the physician shall do no harm, but unfortunately the relative nature of this harm when put in context with terminal patients has been perverted. The balance between regulation and doing no harm must be judged by the stakes involved, and most importantly the patient and physician must be put back into the driver's seat regarding life or death treatment decisions, not a regulatory agency. As we have seen in Chapter 6, all drugs harm some patients, even aspirin. Existing decades-old chemotherapy drugs are extremely toxic, they kill all cells, and so what are we really talking about anyway? Can the new investigational drug candidates be any worse? This, coupled with the variation in patient response to all drugs that has been discussed at length already in Chapter 6, mitigates strongly for looser guidelines in accessing the risk–benefit equation employed in the drug approval process and the availability of investigational drugs. The FDA is a big part of the reason that terminal diseases still plague us and Washington can change it quickly and easily with little or no cost.

Before discussing how the FDA needs to change, it is important to understand the power that the agency has in controlling patient access to drugs. While some believe that the FDA is in the pocket of the pharmaceutical industry, I can testify from personal experience that nothing could be more untrue. The agency has total sway over the fate of all drugs and devices, and CEO of every drug company knows that. While this is somewhat justifiable when it comes to drugs that are variants on pain killers or blood pressure drugs, it is deadly in terminal diseases and rises to the level of a crime against a patient's right to life. When it comes to terminal disease, we need a totally new regulatory approach, a new FDA with a different risk–benefit perspective and a more robust commitment to both the patient's right to life and to act much more quickly than it currently does. It must be aggressive to a degree consistent with the gravity of the crisis to humanity that the terminal diseases constitute and committed to work in partnership with the drug industry, physician, patient, and the research community rather than its current adversarial role.

RETHINKING THE FDA

The FDA is overwhelmed and grossly underfunded, and this will not be cured by splitting off the regulation of foods, which should have been done decades ago. This stems from the history of the FDA, which was initially to assure that new drugs were safe, that is a very simple task. It was only later that establishing the efficacy of new drugs was added to the mission of the agency, and unfortunately it was not accompanied by the funding and staffing required for that and the enormous explosion of start-up biotech companies since the mid-1970s. Proof of efficacy is a much more difficult task, and it requires judgment and a different kind of leadership and staffing. We have already discussed how different each patient's response is to a given drug and how human clinical trials differ from those conducted in animals that are either isogenic twins or at least of the same breed if done in larger animals. While this is a problem with all drug clinical trials, it becomes a much more severe problem when the drug in question is a patient's only alternative to certain death. The current FDA does not seem to realize that and continues to have the same statistical approval parameters ($P < .05$) that are employed for me-too drugs.

The FDA treats terminal disease drugs in a manner not sufficiently different from me-too drugs. Terminal disease drug approvals basically require the same arduous approval process, the same strict reliance on statistics, and in most cases the same requirement on placebo-controlled trials. This takes too long and costs too many lives. Terminal disease drugs wait in the same cue as me-too drugs and require the same three phases of clinical trials before dying patients have the right to try them. Often the time required to demonstrate sufficient Phase III results for a promising Phase II drug can be as much as 5 years. This makes no sense when the patients' only other option is entering a hospice center. The access to promising terminal drugs needs to be dramatically accelerated in a controlled setting where the first several hundred patients treated are closely monitored. For that to happen, three things need to change in the psychology of the FDA. They are the risk–benefit equation, the primacy of a terminal patient's right to life, and trust in the physician–patient relationship. When the benefit of being exposed to an experimental drug is possible extension of life or a remission, then more risk can be tolerated. The patient in collaboration with their physician should be able to decide whether they want to try an experimental drug that has shown encouraging results in much smaller numbers of subjects than is required for formal FDA approval. This is currently very difficult to do and has to change.

There also needs to be more communication between the FDA and the drug sponsors. The relationship of the FDA and the sponsor should be more of a two-way discussion or a partnership rather than a stiff formal one with limited in-person contact. More and a different kind of staff with industry experience is needed to accomplish this goal. All drug companies fear the FDA and routinely wonder what they are thinking regarding their drug. This leads to false starts and wasted time and that costs lives when the disease in question is terminal. More face-to-face communication would greatly improve efficiency and remove a lot of the mystery involved in getting a drug approved or taking it off the table.

FDA reviewers need to get closer to the drugs they review rather than simply read reams of documents once every year or so and have to refresh their memories of what transpired at previous meetings. Meeting minutes are useful, but they are no substitute for in-person contacts. Regulators of terminal disease drugs need to be part of the drug development process for terminal disease drugs. This includes not only safety and efficacy studies but also the manufacturing and QA/QC aspects. This would reduce errors in communications and expectations, which are a major cause of drug approval delays. It would also allow regulators to give the sponsor's greater feedback on the development and treatment rationale for the drug in question and offer more comments on issues such as clinical trial design, manufacturing, and risk–benefit analysis. Too often these issues surface very late in the drug development process, and correcting them requires starting a 3–5 year process all over again. While this is unfortunate when drugs for nonterminal diseases are under review, it is catastrophic with terminal disease drugs.

The nature of the staff also needs to change. Many drug reviewers have no experience in drug development. They are generally PhDs with chemistry or biology expertise or MDs with academic backgrounds and very little clinical experience. More senior people with experience in the specific disciplines of drug development (medicinal chemistry, toxicology, clinical medicine, manufacturing, etc.) are needed. A good scientist is often a bad drug reviewer, and this can delay or destroy a good drug in development. The division directors at the FDA are mostly MDs with very limited management or drug industry experience and have to be replaced by seasoned managers with drug industry experience.

Many additional burdens have been placed on the FDA over the past several decades, and they have not been accompanied by the appropriate additional resources. The Fast-Track or Expedited Review initiatives of the 1980s and 1990s that resulted in the rapid approval of several new AIDS drugs are a good example. They worked but were not accompanied by the resources required to perpetuate and expand them to other terminal diseases. AIDS is now a chronic treatable condition rather than the death sentence it was before these programs. These efforts are important and can make a difference as the AIDS example clearly demonstrated and must be invigorated rather than reigned in. A more rationally staffed and funded FDA will save the healthcare system billions of dollars per year if it results in more rapid judgments on drug approvals and tens of billions per year if these accelerated judgments result in saved lives and expedited cures. The cost of getting a new drug approved has mushroomed to over a billion dollars and often takes over 10 years to complete, and it is totally due to the FDA. This is lunacy.

So there are many fundamental changes to the existing FDA that would make it quicker and more efficient in getting terminal disease drugs to patients. The question is whether these changes should occur under the auspices of the current FDA or should they go a step further and split off terminal disease regulation as suggested in the previous chapter. Before discussing specifics of that, it is important to make the case just how haphazard and inefficient the FDA management teams have been over the years, and the best way to do this is with real-life examples.

SOME COMMON FDA HORROR STORIES

Every CEO has stories about how much the FDA has delayed the approval of their drug candidates and the resultant cost to their company. Most of these are pure management foul-ups and a maze of regulations that derive from their distrust of the drug industry. The agency also has no real accountability for the slow response and foul-ups it imposes on drug sponsors. It does not monitor internal staff sufficiently and exert the discipline required to correct these problems that can in some cases kill companies. Here are just some of my frustrations during the decades I have dealt directly with the agency.

One of my companies did three Phase II clinical trials with a drug for three different manifestations of an acute trauma to the heart with the goal of reducing the damage to the heart. The rationale behind this was to determine which trauma scenario responded best to the new drug we were trying to get approved. When it was time to start the Phase III clinical trial, we chose one of those indications to move forward with and informed the appropriate division of the FDA we were working with about our plans. They said fine and advised us to move forward with the preparations for the start of Phase III meeting with them. We sent an outline of the trial design with the endpoints during these conversations and proceeded in a 9-month preparation for the meeting. This is an enormous task, which required us to hire numerous consultants; identify, validate, and recruit dozens of clinical trial sites; and incur numerous other expenses. We informed our shareholders of this with a press release since we were a public company and were obliged to do that. We then flew a large team to the meeting with the FDA only to be greeted by the head of the entire drug component of the FDA and our division director informing us that we would have to refile the entire drug INDA with a different division of the FDA. We were stunned because we had fully informed our division director of our plans 9 months before. But the director of the entire FDA felt it would fit better if we switched to another division. Well why did they not say so 9 months earlier? It stunned all of our internal and external experts who advised against engaging the ombudsman, which was my immediate thought.

It took us another 6 months to get up to speed with an entirely new FDA division that knew nothing about the specific disease we were studying and had a whole different set of issues, which we had to deal with. The bottom line is that we lost 15 months and about $12 million in investor capital to get the Phase III trial started. One year later we were forced to stop the trial due to lack of funds because our investors could not provide the additional funds required to complete the trial. All of this was due to management errors of the FDA. This potentially breakthrough therapeutic approach to a very poorly treated acute cardiovascular crisis was never developed. No one was fired at the FDA or even disciplined to our knowledge. If a company caused a consumer such grief and expense, it would have been sued to the extent of bankruptcy, its reputation ruined, the CEO and executive staff fired, but for the FDA it is just another day at the office. Now for story number 2.

Ten years ago I was CEO of another company that had a novel drug to treat a metabolic disorder in newborn infants. It had been widely vetted in hundreds of patients in investigator-sponsored clinical studies and the results published in numerous journal articles. The FDA allowed us to start a 300-patient Phase III trial with this drug in newborn infants to prevent this metabolic disorder from occurring. We went through all the rigorous preparations for starting such a trial with the FDA's blessing. Nine months into the trial with over 100 patients treated and absolutely no adverse events attributed to the drug, we get a letter from the FDA telling us to stop the trial because the FDA decided it wanted to convene an Advisory Committee meeting to access the drug because it was such a new approach. This is a huge event as discussed earlier in Chapter 5 that took us 6 months to prepare for and took 2 full days of presentations. Scores of experts were there attesting to the value of the drug including the then director of the NIH institute that dealt with this disorder who recommended that every newborn baby should get this drug to prevent the disorder. To make a long and agonizing story short, 1 month later FDA told us to redesign the trial with a totally different population of infants. This voided the previous trial and would have required us to start an entirely new trial, which would have required 9 months to start. The company could not do that for lack of funds and had to reorganize and fire most of the staff and search for new investors. The drug is still in limbo; again, a catastrophe for the company and just another day at the office for the FDA.

I could go on but my stories are tame by comparison to others that I have heard from CEOs who do not want to publicize them. They all speak to an agency that lacks the organizational skills required to interface with industry where timelines and budgets are required. This, coupled with a bureaucratic mentality built on tenure, academic purism, and the lack of accountability to the patients they serve, is a serious impediment to the entire drug development process.

Possibly the management issues are just the result of staff shortages, but that can be remedied if the directors were so motivated. All federal agencies are inefficient due to the very nature of the tasks involved. The stakes with the FDA are, however, of such a nature that it cannot be tolerated. FDA has the added problem in that it has developed an adversarial relationship with the industry it regulates. For every year a cancer cure is delayed about 600,000 lives are lost; for cardiovascular disease another 800,000 are lost; and for neurodegenerative disease several hundred thousand more; yet we all accept this absolute power that the FDA thrusts on the entities charged with providing breakthrough drugs. You would think the fact that only 22 new drugs have been approved by the FDA in all of 2016 would have some consequences; it is long overdue to make that happen.

ACADEMIC PURISM VERSUS RIGHT TO LIFE

The examples of situations where the FDA has delayed important drugs because of minor details and pure academic issues are legion, such as holding up a potential lifesaving drug because it just wants another trial to be sure of the statistics or requiring a placebo control when doing such would be unethical or extremely difficult [4]. The agency, however, reigns supreme in all these instances with absolutely no recourse for patients, physicians, or even Congress. Patients, their families, and the sponsoring companies are forced to suffer in silence. There either has to be a law passed or an amendment to the constitution to rectify this fundamental violation to a patient's right to life. The "right-to-try" movement that is gaining attention would enable terminal patients to try investigational drugs, but this is not enough since once enacted it will probably be eviscerated in the enforcement arena by the FDA as other accelerated review initiatives have.

Another drastic action that would stop the FDA in its tracks is to revert back to the pre-1963 agency and restrict the mandate conferred by Congress to only safety and not efficacy as some have suggested [11,12]. This is probably overkill, and a more rational approach would involve greatly lowering the bar on the efficacy standard for approval to post-Phase II when it comes to drugs that treat the terminal diseases. This would enable patients to make their own life and death decisions in concert with their physicians and the sponsoring drug company. Such an initiative would greatly reduce the cost of terminal drug development and the time it takes to get promising drugs to patients dramatically. It would also act as a huge incentive for innovative start-up companies to focus on terminal disease drugs rather than me-too drugs.

There will be those who argue against this reduction in FDA power over our lives, but the stakes are such that whatever the false starts might be in making investigational drugs available to patients, the reality is that the current system of approving these drugs has failed and the degree of regulatory reach into the most fundamental human right is not consistent with a free society. Some form of efficacy should be required before drugs are made available to patients, but it must take into account the need to speed up the cure quest and offer death sentence patients hope.

PLACEBO-CONTROLLED TRIALS FOR TERMINAL DISEASE DRUGS

The need for placebo-controlled trials for terminal disease drugs is a prime example of academic purism trumping common sense. The most recent example of this is the fiasco played out around the muscular dystrophy drug eteplirsen [4]. Numerous experts in the field petitioned the FDA to approve this drug for kids who suffer from this disease, which cripples them at a rapid rate resulting in death. The experts based their plea on clinical data showing a significant reduction in the progression rate of the disease based on historical controls and not placebo controls. The drug is based on an entirely different approach to the disease, which appears to be safe and well tolerated. The kids have no other alternative other than death. The FDA is, however, insisting that additional clinical trials be performed using placebo controls before it will approve it. The trials will take several years, and this is with full knowledge of the fact that it is not possible to repair existing muscle damage once it occurs in muscular dystrophy but rather only to slow or prevent future damage. The FDA is playing God over these kids in direct opposition to parents and expert physicians. They are taking the right to life from these kids, condemning the parents to a no-hope scenario, and insulting the treating physician. Eventually the FDA succumbed to intense pressure and allowed the drug to be used if a placebo-controlled trial was started.

The entire idea of a placebo control is unethical in a terminal disease, yet the FDA has an uncontested mandate to do this! How does this happen and what will it take to remove this ridiculous trampling of ethics, basic human rights, and simple common sense. Placebo-controlled trials in terminal disease must be stopped. Historical controls coupled with a lack of serious side effects should be sufficient to warrant approval and then let physicians and patients decide whether the drug is worth using. Congress has to get involved here and that will only happen if citizens make this a high priority issue like the gay community did over the AIDS situation.

A REGULATORY–INDUSTRY–RESEARCH PARTNERSHIP

So what is the bottom line for fixing this evolved monster of an FDA? What appears obvious is that past outcries to tweak the FDA's grip on terminal disease drugs have met with some reforms such as Fast-Track status and compassionate use provisions, but they soon become buried in the agency and either made too cumbersome to use or simply ignored, they have not made a difference. The regulation of terminal disease drugs has to be entirely rethought and moved out of the FDA with an entirely different management team and staff. Regulatory bodies only grow in power and reforming them is difficult, especially when it is only a part of the agency that needs attention. Strict regulation is appropriate for me-too drugs for which a case can be made that they should not even be generated. They serve the drug industry by generating life extension strategies for drugs that are going off patent but pale in importance when compared to the drugs that are really needed.

In Chapter 7 where reforms for the NIH were discussed, mention was made about the concept of a partnership between regulators, the drug industry, and the research arena when it comes to terminal diseases. In the NIH/TDA discussion it was suggested that regulators actually work directly with the drug developers and researchers to speed the access of investigational drugs to terminal patients and also greatly reduce the cost of getting these drugs into patients. Terminal disease regulatory functions would be housed in the TDA and under its jurisdiction, not the FDA. This would revolutionize drug development. Such a relationship between government and industry has proven successful in many of the transformative technological advances made over the past 50 years. This type of cooperation, organization, and urgency where government and industry work together is sorely needed to find cures.

The regulation of terminal disease drugs would no longer be conducted within the FDA but would rather become the fourth institute within the newly formed TDA. This will enable the regulators of terminal disease drugs to work more as partners with the drug discovery and development people as well as the physician and patient under the auspices of the TDA. The regulatory staff would have real-time access to all the drug development data internally developed at the TDA, and the joint mission would be getting new therapies into patients as soon as possible. Translational medicine, that is the translation of research into a product, which has now become a buzz word in medical research circles that everyone is trying to achieve, will take on a new meaning in such an environment since it will be a one-step process from the bench to the clinic all happening under one roof. Drug companies would communicate with the regulators through the TDA in such a manner that would enable a more coordinated and strategic approach to drug development. The most promising drug candidates either generated within the TDA or from drug companies could be candidates for TDA support through the expanded and upgraded clinical facilities that would be in place.

The TDA would become a true partner with the regulators and industry serving as a focal point for the entire effort involved in the development of cures. This would be a quantum step forward that would directly remedy the lack of coordination that makes the current R&D establishment so chaotic. This centralization of the entire cure effort would be managed by project leaders who would deal with both industry partners who would have on site representation and intramural TDA staff who would all be required to focus all their efforts directly on the disease rather than follow their own academic interests. There would be more focus on applied research involving innovative exploitation of existing science, the repurposing of existing drugs, and combinations thereof to the three terminal disease classes. Basic research involving new frontiers in biomedical science would continue but in the context that it would be long term and not sold to the public as the only hope for the near-term cure they are not yet prepared to provide. The regulators would be exposed to all of it and be facilitators rather than adversaries.

The larger drug companies should also be willing to support some of the efforts of the TDA with regard to their own investigational drug candidates. Certainly this would be profitable to the drug companies given the shorter development timelines and reduced expenses that would ensue from the efficiencies of such a partnership. If such an arrangement would reduce the average development cost of a drug by several hundred million dollars for a company than a substantial investment, in the form of a fee, in TDA, infrastructure support would be justified from a bottom-line perspective. The practicalities of such an arrangement need to be formalized but it is at least in concept, an improvement over the current adversary relationship between the FDA and industry. Think of it; no need for the translational medicine two step and the industry paying a chunk of the TDA budget, and drugs coming to patients in one third the time.

This industry–regulatory–research partnership would revolutionize the efficiency of the terminal disease cure effort. The patient would finally become the epicenter of the mission rather than a helpless observer. For the first time the major components of the effort would be on the same side of the table rather than adversaries.

Regarding the regulatory function for terminal disease drugs, each terminal disease category would constitute a branch of this new TDA regulatory institute; one for neurodegenerative disorders, another for cancer, and a third for cardiovascular disease. This is the first step in the restructuring of the terminal disease regulatory effort, remove it from the FDA; staff it with proven managers; and partner it with the R&D people, but more is required. Patient access to new drugs has to be simplified and accelerated. Three stages of clinical trials spanning almost a decade are absolutely ridiculous when it comes to these drugs. This dictates a snail's pace to the development pipeline and an enormous burden to the companies, especially the small innovative start-up companies. Instead of three clinical development stages there will only be two and they will be shortened. There will be a 32-patient Phase I safety trial performed at the TDA and a 90-patient Phase II trial at two different drug doses of the study drug. No placebo group would be used in any of the trials, but rather a group of patients would be included who are treated with the best available approved drug or drug cocktail for that disease. This would select for a significant superiority of the new drug relative to the standard of care. This clinical trial scenario will greatly shorten the patient access time and cost. Four to six years could be chopped off the development cycle, and the costs reduced by several fold since the most expensive component of clinical development are the larger Phase II trials now employed and the two multicenter Phase III trials. The through-put of cure candidates would be greatly increased. Also the statistical requirements for gaining access to patients would be reduced to a $P<.10$ rather than the current $P<.05$ and hold for both primary and secondary endpoints.

Lightening the regulatory component of terminal disease drugs would not only get more drugs into patients faster but it would also act as an incentive for industry to reduce their emphasis on me-too drugs and focus on cures. Why confront a 10- to 12-year time to product when you could chose a 4- to 6-year product access period. It would incentivize innovation, provide much needed hope for patients, and get physicians more actively involved in the research effort. A key component of such an accelerated development cycle for terminal disease drugs would be postmarketing surveillance of the clinical response and toxicity profile of the new drug. The treating physician, drug sponsor, and regulatory staff would be required to provide and document such data and then render a judgment as to whether the drug should be formally approved for all patients, only a subgroup, or not at all.

The current FDA is intoxicated with power that can just barely be tolerated for regulating me-too drugs; the record is clear on this. The terminal diseases require undivided attention and an urgency that the FDA has not been able to provide. It requires a proximity and partnership between industry, researchers, physicians, and regulators that currently does not exist. The suggestions for reform presented herein may prove threatening to some of the current stakeholders but it nevertheless needs to be discussed to stop the strangulation of the cure enterprise and to generate a sharply focused mission.

REFERENCES

[1] J. Whalen, Hurdles multiply for latest drugs, The Wall Street Journal (August 1, 2011). Retrieved from: www.wsj.com/articles/SB10001424053111904233404576459851152423110.

[2] A. Von Eschenbach, Toward a 21st-century FDA, The Wall Street Journal (April 15, 2012). Retrieved from: www.wsj.com/articles/SB10001424052702303815404577331673917964962.

[3] R. Goldberg, Government Is Stifling Medical Innovation, The San Diego Union Tribune, March 13, 2011. Retrieved from: www.sandiegouniontribune.com/opinion/commentary/sdut-government-is-stifling-medical-innovation-2011mar13-story.html.

[4] The FDA vs. Austin Leclaire, The Wall Street Journal (April 22, 2016). Retrieved from: www.wsj.com/articles/the-fda-vs-austin-leclaire-1461281386.

[5] The real FDA scandal, The Wall Street Journal (February 6, 2008) A18.

[6] The FDA and slower cures, The Wall Street Journal (February 28, 2011) A18.

[7] Where's the drug, FDA? The Wall Street Journal (July 2, 2016). Retrieved from: www.wsj.com/articles/wheres-the-drug-fda-1467413266.

[8] The Abigail Alliance. Retrieved from: http://www.abigail-alliance.org/story.php#.

[9] R. Nelson, "Right to Try" bill in Senate for terminally ill patients, Medscape Magazine (November 20, 2016). Retrieved from: http://www.medscape.com/viewarticle/863336.

[10] B.A. Cohen-Kurzrock, P.R. Cohen, R. Kurzrock, Health policy: the right to try is embodied in the right to die, Nature Reviews Clinical Oncology 13 (May 24, 2016) 399–400, http://dx.doi.org/10.1038/nrclinonc.2016.73.

[11] M. Boldrin, S.J. Swamidas, A new bargain for drug approvals, The Wall Street Journal (July 27, 2011). Retrieved from: www.wsj.com/articles/SB10001424052702303812104576441610360466984.

[12] B. Frist, T. Coburn, Streamlining medicine and saving lives, The Wall Street Journal (May 12, 2016) A15.

Chapter 9

Enabling the Drug Industry

Chapter Outline

The drug industry is like no other business. The roadblocks to marketing a successful product are enormous. Why would one want to put capital at risk to develop a product that must be approved by a cumbersome government agency that does not trust them, defers to academics, requires 10+ years of high risk extremely costly development work, and exposes them to enormous potential liability during the entire product lifecycle? It borders on suicidal. If the science does not get you, the Food and Drug Administration (FDA) will or the capital will run out or the lawyers will destroy you. Much of the drug industries failures to cure terminal disease stems from outside forces such as the FDA, patent policy, and medical Tort law, but first let us focus on what the industry itself is doing wrong.

TREATING IS MORE PROFITABLE THAN CURING

A successful business must have recurring revenue. Treating a chronic disease means that the patient is a customer for extended periods and that means cash flow, which is a main motivator of the drug or any industry. The vaccine business is a good example. Most companies are not interested in a drug that requires only one treatment every 5 or 10 years or more. The clinical trials are very difficult to do because the patients generally have to be monitored for many years to show efficacy. In addition and perhaps more troubling is the fact that the drug is only given once. This is somewhat mitigated by the fact that large numbers of people would take the vaccine, but still, the revenue stream is generally not as attractive as a chronic disease drug.

Under the current status quo it makes no sense for drug companies to pursue robust cures. Treating a disease from many different angles that increasingly

A Roadmap for Curing Cancer, Alzheimer's, and Cardiovascular Disease.
http://dx.doi.org/10.1016/B978-0-12-812796-4.00009-X

hold it at bay is a much safer and more profitable approach. The patient is a customer for life rather than for a moment. Three things are required before industry even considers changing their focus from a treatment mentality to finding cures. First and foremost there must be financial incentives. Second, the regulatory (FDA) environment must make it easier to get such potential curative drugs into clinical practice sooner [the terminal disease administration (TDA) will take care of much of that]. Third, and not least, the legal system must be reined in regarding drug liability and medical malpractice. Until these issues are addressed, it is unlikely that industry will do anything other than pay slip service to developing cures. They will protect their existing revenue streams and franchises at all costs in the current environment.

ENDING ME-TOO DRUGS

Having already discussed what me-too drugs are, it is now appropriate to talk about what needs to be done about them. It is very simple; the FDA should not even accept new drug applications for them anymore. Until the terminal diseases are no longer a threat, the FDA should simply ignore them. A list of such therapeutic areas should be developed by the FDA and agreed on by the practicing medical profession. Some examples that might be considered are antihypertensives (blood pressure), analgesics (pain), cholesterol-lowering drugs, and oral antidiabetic drugs. This would force drug companies to devote more serious attention to more important and less well-served areas.

To some extent managed care and the other payers are already discouraging the development of newer me-too drugs by denying reimbursement for their use since many of them are not generic and cost prohibitive. Also the advent of evidence-based outcomes analysis is playing an important role in reducing the use of some of the newer and more expensive me-too drugs. One good example of this is in the antihypertensive arena. One of the first antihypertensive drugs was the diuretic thiazides. They work very well and have been generic and very inexpensive for many years. They were followed by beta blockers, calcium channel blockers, and acetylcholinesterase (ACE) inhibitors, all of which exist in multiple variations. Each type of drug is sold by many different companies, i.e., there are many different calcium channel blockers, beta blockers, and ACE inhibitors. So we have four different classes of antihypertensives that exist in multiple me-too variations and therefore, in a sense, we have me-toos of me-toos, an absolutely ridiculous situation. But for multiple companies these drugs bring in billions of dollars in high-margin revenue. This me-too fiasco diverts enormous resources from the real issues in healthcare and it must stop.

MAVERICKS ARE NEEDED IN THE INDUSTRY TOO

The drug industry spends billions of dollars per year on research. In spite of this they are constantly plagued by patent expirations and are forced to seek new technologies from start-up venture capital (VC) companies through corporate partnerships. Innovation seems to be a very rare commodity at the big pharmaceutical companies. Why is this so? The money is there, and the landscape is littered with PhDs and MDs eager to do medical research. Industry pays much better than academia and has more capital to fund research without the bother of getting grants, so it should literally pump out innovative new drugs that could cure rather than treat the killer diseases.

Perhaps the reason why the innovation is not forthcoming resides in the mentality of the organization and the type of scientists it recruits. Research directors in companies are typically academic scientists or research physicians and this is the problem. The academic mind-set is to play inside the science box and find new science that might have an application. The newest science is always considered the best and is expected to have the best chance of leading to breakthrough treatments and possibly cures. The rationale is that with older science everyone must have already written it off as a therapeutic. The problem is that after a half century of this approach, the cures have not come. Well, maybe it just needs more time? Maybe we should push harder at biotechnology or gene therapy or stem cell research? Certainly these research areas should not be dropped, but counting on them to produce cures in the next 10 years is a very high-risk gamble that has failed multiple times in the past half century.

The industry needs to infuse their research effort with mavericks and innovators and not just thought leaders and Nobel Laureates. The entire body of biochemical and medical science must be reexamined in the context of the three terminal diseases and potential short-term cures. At least 50% of pharmaceutical research should be directed at this extensive mining of existing science and approved drugs. The industry has dismissed older science under the assumption that there are no opportunities left there and this is simply not true. Old drugs can be used for new diseases as we have seen. Metabolic pathways that were discovered and studied decades ago should be therapeutic targets for new drug development, yet the brightest minds in academia are being poured into the newest cutting edge science prematurely. Applied research targeted at the enormous body of existing medical science literature and the thousands of marketed drugs that are off patent must be greatly expanded and encouraged, even if the science is decades old. Certainly the cutting edge science should not be dropped, but the ratio of applied to basic science must increase dramatically and it should all be aimed at the terminal diseases.

If the FDA refused to review me-too drugs, pharmaceutical companies might be forced to look for new uses of old drugs as potential cures of terminal diseases much like the common antibiotics were found to be for peptic ulcers, or thalidomide for leprosy and now a host of other diseases, or aspirin for embolic heart disease. All of these advances occurred without a proactive effort, imagine what might happen if we actually encouraged maverick innovators to mine existing vetted science, look at it in a new way, and apply it for cures. The stigma attached to older science must be eradicated. This will require changes in patent law that would make it possible to file new patents for old drugs and older science when it is used in novel ways to treat terminal diseases. The patenting of older science is necessary to make the capital infusions required to develop these new applications. This can be done now, but the problem is that these patents are rarely allowed by the patent office and are difficult to defend in court. That can be changed as we will see in the next chapter.

The goal is to turn over a lot more stones than we currently are and to look at old science in new ways rather than dismiss it as irrelevant. Perhaps even combining nutritional and herbal therapy with known drugs? There are many potential strategies that are simply not being tried, and these are where the near-term potential breakthroughs will come from. Companies must elevate innovation and maverick science to the level of cutting edge science. The unleashing of innovation and the encouragement of trying things that thought leaders think will not work is a quick way to get answers, especially when there is a long history of trying things that should work and did not. More varieties of thinking need to come into the effort.

Why is it that some people never get cancer or Alzheimer's or have clear arteries into their 80s? This needs to be answered and maybe therein lays a cure. Certain races and geographic regions are less likely to get certain diseases; that needs to be studied more. Academics shun these approaches because they do not involve cutting edge science but rather imagination, innovation, and using ones training in nontraditional ways.

The changes that need to occur in the traditional pharmaceutical industry are dramatic. They need to drop the me-too strategy, focus more on cures rather than treatment, greatly expand innovation around the enormous body of existing science, and expand efforts to find new uses for the thousands of existing drugs and combinations thereof. This will serve to address the critical goal of finding near-term cures, and this is what is currently lacking in the pharmaceutical industry. By all means continue the cutting edge research but scale it to a more reasonable level and do not put all your hopes in that very tenuous arena. There is universal agreement that virtually all the new science areas such as stem cell, gene therapy, and genetic engineering will probably take decades to produce cures, and from a human life and financial cost perspective, it is entirely unacceptable to simply depend on it.

THE VENTURE CAPITAL FIASCO

The VC-financed sector of the pharmaceutical and biotechnology industry has exploded since the 1970s and resulted in thousands of new companies. Many are biotechnology companies led and populated by academic scientists trying to commercialize their basic research work that originated at a university or research institute. Most of these venture-backed companies end up failing or reinventing themselves by licensing in a drug candidate from a larger pharmaceutical firm or buying some older drugs to market and generate a revenue stream. The public gets an inflated impression of what these companies contribute to curing disease. They view them as the last hope for cures and that the new science they attempt to exploit will eclipse the established pharmaceutical industry. After 40 years of the "new medical VC industry" the accomplishments achieved for the hundreds of billions spent is not impressive at best and probably more accurately characterized as poor in general and virtually nonexistent as far as cures for the major terminal diseases.

As discussed in previous chapters the primary accomplishment of biotechnology, an industry that was spawned by VC, has been a small crop of very expensive drugs that have had marginal effects on terminal disease treatment. Many of the companies have made money for the VCs, but far fewer have for the public market investors and only a handful have ever achieved profitability. Precious few have even come close to finding a cure in spite of the fanfare and expectations planted in the public consciousness by a media hungry for news. Google "Venture Capital/Disease Cures" and all you will find is a bunch of statistics on how much VC has been spent on different disease classes and virtually nothing about cures. Why has the VC-backed start-up biopharmaceutical industry been such a failure?

A primary reason for the VC fiasco is the VCs themselves. Very few understand what they are funding, and they rely totally on outside consultants who are largely the thought leaders we have spoken so much about. These thought leaders are only too anxious to get money and attention behind their ideas no matter how fragmentary the evidence is for their commercial utility. Often the VC will opt for preliminary breakthrough science with ironclad patents because they know that they will be long gone from the investment before it ever gets close to the type of scrutiny that will determine its product potential. They preach otherwise, but the reality is that few VCs are involved for the long haul with medical companies like they often are for high technology companies where the product development cycles are much shorter and lower in risk. New wave thought leader blessed science achieves the purpose of getting them to an exit point and then they leave the marketplace to sort out how to salvage the company when the technology fails.

In the early days of VC-funded biotech companies, all one needed was a thought leader founder, a Nobel Laureate on the advisory board, and a patent application to get funded. Large sums of money were wasted on building laboratories, hiring inflated staffs, and convincing the world that this brand new company is all of a sudden a "leader" in a particular field of drug development when they are still a decade away from their first approved product.

VC is a business, and its practitioners should not be blamed for the motivations regarding their investments, after all it is a free country. Of course they would rather have one of their medical companies develop a cure for cancer, but it does not constitute a major reason for making an investment. The exit strategy is the key, and it is timed in a manner that serves the investment and not the potential product. The VC investment strategy often forces the company to do things such as a public offering of stock or merge with another firm. Most VC firms raise money on a 4–6 year cycle, so they must show a return to their limited partners during that time frame to generate the next capital raise or fund. This nonproduct centric approach of VCs is a major impediment to the product flow from venture-backed companies. It is all about timing entry and exit points and not always what is best for the company.

The venture-backed segment of the drug business therefore suffers from a different set of problems than the traditional pharmaceutical industry. Rather than an infatuation with me-too drugs and safe product bets, the venture-backed industry is too "virgin science" oriented and not concerned enough about whether an actual product emerges from their companies. They most often create an expectation and feed off the insatiable opportunism of the investing public to get in on the ground floor of the newest best thing. The scenario is getting very old and the public is beginning to wake up to the rouse, but the reality is that the cause of finding cures has not been well served by the VC industry. The timelines are simply too long for VCs, and the variability in the science creates extremely high risks. The science, expense, time, and regulatory risk are beyond their capabilities, so they have become masters of building near-term development potential and then exit before judgment day.

Another significant problem with VC companies is their almost universal emasculation of the entrepreneurial spirit. They typically own a majority stake in their companies and in so doing convert entrepreneurs into employees. Scientific founders often need extensive help in translating science into products, but there is a right and wrong way to do this. Taking too much of their company is not the right way, but again business is their motivation, not cures. Often there will be multiple rounds of funding for a venture-backed company. Extensive syndicates of as many as a dozen or more VC firms are often investors in a single company, and board meetings require name tags to remind management who is who. This creates anything but an entrepreneurial culture and inhibits innovation. VC corporate boards are like no others. They absolutely control the companies they fund with the CEO serving simply as a spokesperson for them. Imagine a large committee each with their own ulterior motives running anything, especially a company with such a tortured product development cycle.

The solutions for the venture-backed part of the industry are somewhat more elusive than for the traditional pharmaceutical industry. Certainly VCs should change their strategy. Relying less on thought leaders and more on industry professionals would help them make more rational choices for funding and some of them do. Being more rational about the science they invest in would also be a step in the right direction. If more entrepreneurs and industry veterans were general partners, this would also help the strategy and efficiency of their companies. Coordination among the different venture firms regarding what they fund would also help. Much like the granting agencies, the right hand often does not know what the left is doing, and a lot of money gets wasted in redundancy. Cooperation is going to be difficult to achieve because VCs compete with each other, but it can be done through syndication rather than taking the lead in the investment.

Perhaps the biggest problem with today's VC industry is that true innovators and mavericks often do not impress them since they do not fit the mold of being members of the mainstream academic community and VCs have no means of dealing with that. The thought leaders and Nobel Laureates they seek counsel from cannot validate these mavericks and innovators since they are not their disciples. This is unfortunate since mavericks and innovators are precisely what are needed to solve problems that resist traditional approaches. The maverick examples discussed earlier clearly show that they are being suppressed by a system that discounts them. VC should flow to such individuals, and VCs need to get better at finding and vetting such scientists rather than following the dictums of a select few "eminent" scientists who have thus far lead us in too few directions to be successful and have clearly failed us.

DO START-UPS MAKE SENSE?

Regarding the small emerging companies in general, one needs to ask whether these VC-funded vehicles to new drugs make sense. The major shortcomings are the academic technology focused mentality, wasteful redundant infrastructure, and most importantly, the singularity of product focus. Small biopharmaceutical companies typically have one major product candidate and several other back up opportunities, but the reality is that success resides in getting their most advanced product candidate approved or partnered. If that does not happen, the show is over. This spawns a make-or-break imperative. If the Phase I or II clinical data looks marginal, these companies have little choice but to move ahead since their other product candidates are so far behind their lead product. Big pharma companies generally have multiple product candidates coming out of Phase II, and they chose the best few to move into Phase III and even they fail more often than they succeed. This lack of multiple opportunities forces small start-ups to move ahead with marginal Phase II data into extremely expensive Phase III trials that are less likely to work and this results in wasted capital, time, and talent.

It is not uncommon for start-up companies to build staffs that number in the hundreds and occupy huge upscale facilities with state-of-the-art labs. All this exists without any source of income other than investor capital. This wasteful extravagance of infrastructure greatly restricts the number of opportunities that such companies can pursue since they often spend several millions of dollars per month just to maintain their staff and infrastructure. Many start-up companies look like research institutes or billion dollar companies. There should be a rule that a company should not exceed 15 people, this alone would greatly improve productivity and, if coupled with outsourcing, could enable moving multiple drug development opportunities forward, and of equal importance, making it easier to terminate projects when the data starts to look bad.

The VC life science model has been more hype than substance. Enormous amounts of capital could have been better used if it were more directed at strategic R&D rather than to support redundant infrastructure and overhead. There needs to be more emphasis on products, innovation, maverick science and multiple opportunities, and less on fancy science projects. The focus should be on business models that generate multiple "proof-of-concept" value points for potential products. These opportunities can then be individually partnered with larger companies at the proof of concept stage. This would minimize wasted capital and enable VC to turn over more stones. There also needs to be much more VC used for exploiting existing science and looking at existing vetted science in new ways, after all its about products and not science.

CORPORATE PARTNERSHIPS, WHY BOTHER?

Big pharma and VC-backed start-ups currently exist in a symbiotic relationship. The large companies have had difficulty innovating for reasons we discussed earlier (me-too drugs, treatment vs. cures, etc.), and they try to solve this by mining the small "entrepreneurial" start-up VC-funded companies for licensing opportunities. This scenario makes a lot of sense for the small company since they cannot survive without the support of the bigger companies. But for the large companies it amounts to an admission that they are not doing their jobs right. First of all, the large companies should not need to partner with small inexperienced companies who know far less about the pragmatic aspects of drug development and whose data is of questionable quality. They spend billions of dollars on research each year and should not need to go to start-up companies to get ideas, at least not on the scale they currently do.

The big pharmas could go directly to the source of the ideas they need rather than pay a middleman for having spent hundreds of millions of dollars to prove a concept that they could have done better and for a lot less money. The large companies should set up small business groups to test out new ideas directly from the scientific literature. This would simply involve repeating academic studies under strict quality controls to either throw it out or move forward with it. Corporate partnering adds a middleman to drug development that is highly inefficient from every perspective such as cost, time, and experience.

In addition to saving millions of dollars on the front end by not licensing technology from a start-up, the large companies could do a much better job in the proof-of-principle phase of drug development. When small companies test their drugs, there is a strong bias since it is all they have and it better work. The large companies have more choices and tend to choose the strongest candidates for future testing using more pragmatic metrics and less emotion. The unnecessary process of partnering is also an enormous time waster since the small company has to be financed, find facilities, build out laboratories, and do a hundred other things that an existing company already has in place. The practice of vetting new ideas by seeding start-up companies that are learning as they go, followed by corporate partnering, is not the desired path for solving a problem as elusive and urgent as developing cures. Add to this the fact that the start-ups claim they exist because they detest the cumbersome aspects of large companies and their inability to innovate, and it is obvious that neither party is comfortable working with the other. Why is it that they seek each other out with such intensity? The answer is that the big companies do not currently provide the right atmosphere to innovate with regard to cures and have little financial incentive to do so. Clearly survival forces the little guys to swallow their egos and partner or else face bankruptcy. It is almost laughable, the start-ups are escaping something they eventually have to return to, and the big guys are utilizing something they could do much better themselves.

The low success rate of corporate partnerships should not surprise anyone remotely familiar with the drug business. The big companies mine the thousands of small companies chasing virgin science with the expectation that they have a foot in the door if the technology works but are not ready to bring it in-house until it does. Much of what goes on in start-ups will happen at facilities such as the proposed TDA, and the proposed reformed universities and research institutes that we discussed in Chapter 7, since now they will be generating vetted quality-controlled data. The TDA is a much more efficient and less biased environment for the proof-of-concept studies than start-ups are since the fate of a company is not dependent on the outcome. The TDA would also have in-house regulatory staff and existing facilities that would further facilitate the development process.

So we have created an entire industry of countless small companies that spend billions of dollars per year essentially doing research in newly created redundant infrastructure in a "learn-as-you-go" organization. Precious VC is being wasted in this uncoordinated chaotic process. The big pharma licensee ends up with an inexperienced partner that must put the best face on the licensed product, which more often than not turns out not to work. VC investors are beginning to realize that medical start-up enterprises rarely succeed and that the only way out of such investments is for a corporate partner to come in and make something happen. What makes the partnering situation even more ludicrous is the fact that most often the small "innovative" company that the large company is trying to license from does not have a "cure" to offer but simply a way to produce a marginal improvement over existing treatments.

The bottom line here is that the vanguard VC-funded start-up companies are not very effective. They are expensive, inefficient, and take too long to produce results. The successes of VCs in the biotech sector are almost by accident. In their defense it is really almost impossible to predict success in the drug business in general. As we have discussed the science is nowhere near being exact. The roadblocks of the FDA are extremely difficult to deal with and the patent and legal system as well. The success rate is worse than that of the restaurant business, but the price of entry is orders of magnitude greater. If VCs insist on staying in the life sciences area, it would make more sense if they would do it on a smaller scale. Instead of company builders they should be idea builders by bringing multiple vetted and patented ideas forward and then licensing them to established companies. They have begun to move in this direction by building smaller companies and outsourcing support functions, but they need to go to the next step and not build the company at all. This could be accomplished by funding university or institute research under strict conditions such as quality controls, milestones, and accountability.

INCENTIVIZING INDUSTRY TO FOCUS ON CURES

The drug industry is in bad shape regarding cures. The big companies are merging with each other, and their patents are expiring faster than the new ideas they generate, a clear sign that they are in trouble. They are unable or unwilling to innovate and have been forced by a stringent regulatory maze into the corner of focusing on safe bets such as me-too drugs that provide incremental advances in treating disease rather than cures. The "entrepreneurial" start-up companies are infatuated with their technology, inexperienced in drug development, and slaves of venture capitalists bent on generating an exit strategy. The solutions proposed above for the reform of the drug industry are a start on the road to facilitating cures. They all, however, will not happen if there is no incentive for them to change focus from treating disease to curing it.

These incentives must focus on three elements: financial, regulatory, and legal. Financial incentives could take the form of tax breaks on revenue generated from curative drugs. Also the patents on such drugs could be enforced for 30 years instead of the current 20 years. Regulatory incentives must include shorter product development paths such as only requiring one Phase II trial to register a terminal disease drug. This one reform would cut both the cost and time to a product by more than half and would be a huge incentive for the entire industry to focus on such treatments. Also, a variant of the Waxman Hatch act that provided for a seven tear marketing exclusivity period for drugs that treat orphan diseases [1,2] should be passed. This orphan drug legislation has been a very successful incentive for the development of such drugs that were largely ignored previously. To incentivize cures for terminal disease the exclusivity period would need to be longer, perhaps 12–15 years. It could be designated the Terminal Disease Exclusivity Act or TDEA, and it is further defined in Chapter 10. The TDEA would provide exclusivity for the novel use of anything in the literature and the public domain that is targeted for terminal disease therapeutics or diagnostics. This would stimulate innovation dramatically since it opens up vast amounts of previously unpatentable technology and would incentivize companies to go after cures. Other changes in patent law and the legal system that will move companies away from me-too drugs will be discussed in Chapter 10.

WHERE IS THE CONSUMER?

Managed care, nationalized medicine, and the potential of price controls are also scaring the wits out of the drug industry and threaten to paralyze innovation. Managed care and the Affordable Care Act remove the patient as a consumer of healthcare. This means that the drive from the consumer to the drug industry for better medical products is nonexistent. If medicine becomes a monopoly with one provider of services that controls prices and availability of new drugs, it will further force the drug industry into safe rather than aggressive medicine. This will inhibit cures by dealing yet another blow to innovation and risk taking. Reinvigorating the patient as a healthcare consumer has been tried with Health Savings Accounts, and hopefully this will catch on as it will surely serve to bring healthcare costs down and provide more drive for cures since patients will finally have some skin in the game. Reforms or repeal of the Affordable Healthcare Act will be required to achieve this in an across the board fashion. Until the consumer is actually shopping for healthcare, there will be no competitive pressure applied to the product it is shopping for and therefore none for the industry to produce it. Consumer input has always driven innovation, and it is totally absent when it comes to medical products. This key driving force of innovation must be mobilized and will be more likely to happen if the consumer learns more about the product and how it serves the intended purpose. Hopefully this book provides a start in that process. It is hard to imagine a more important and deserving product for your attention.

The drug industry reforms discussed are intended as a starting point for changes that would help focus it on cures but for them to take place reforms are needed in other parts of the drug development process. They are all interlinked and impinge on the drug industries' ability to function effectively. If we do not unbridal and incentivize the industry, it will continue to limp forward at great cost to humanity.

REFERENCES

[1] C. Helfand, Top 20 orphan drugs by 2018, Fierce Pharma. Retrieved from: http://www.fiercepharma.com/special-report/top-20-orphan-drugs-by-2018.
[2] E. Rensi, The orphan drug act as been a huge success, The Wall Street Journal (June 23, 2008). Retrieved from: www.wsj.com/articles/SB121417838559995535.

FURTHER READING

[1] When the peers all think the same, The Wall Street Journal (August 6, 2014). Retrieved from: www.wsj.com/articles/when-the-peers-all-think-the-same-letters-to-the-editor-1407265925.

Chapter 10

Patent and Liability Reform

Chapter Outline

The final roadblock to cures is the legal system and specifically therein the rules regarding patents and the tort practices regarding liability, class action lawsuits, and physician malpractice lawsuits. The entire legal system is antagonistic to the progress required and if adjustments are not made many of the reforms called for in previous chapters will either not happen or be greatly muted. The R&D establishment has many moving parts, and previous efforts to speed cures have tended to be piecemeal in nature rather than the global across the board rethinking required. For example, if the patent codes are not reformed and the Terminal Disease Exclusivity Act (TDEA) is not passed, the industry will not be able to exploit published science in the public domain and will not be incentivized to spend more time on going after cures rather than pursue me-too drugs. Also if the terminal disease administration is not formed with its new approach to regulating terminal disease drugs, the industry reforms that incentivize cure R&D, i.e., greatly reducing the time and cost for getting terminal disease drugs to market, will be pointless.

SPEEDING UP PATENTS

Drug development requires enormous capital and it makes no sense to engage in it without the assurance that no one will simply copy your drug and walk away with it. Patents are therefore a major activity that all drug companies, researchers, and entrepreneurs must engage in to establish legal ownership of their invention. If patents are not filed and prosecuted with vigor before, during, and after a drug is developed and approved, one should not even bother to start the process. The rule is to build a broad fence around your product or technology, so the competition will not even think of trying to challenge them in court. The test of an intellectual property position is how well it will hold up

A Roadmap for Curing Cancer, Alzheimer's, and Cardiovascular Disease.
http://dx.doi.org/10.1016/B978-0-12-812796-4.00010-6

in court. Many companies spend a lot of time just challenging patents trying to get around them, so they can spin off generic copies of the drug or technology in question. Such cases can drag on for years and divert both capital and attention from more relevant activities.

The road to getting a medical patent allowed and issued is arduous, expensive, and fraught with overworked patent examiners who take forever to respond to inventors. It can take in excess of 5 years to get a patent issued. While this may be acceptable for patents on widgets, it is unacceptable for a start-up medical company trying to raise capital to support the development of new therapies or a large company trying to decide whether to start a Phase III trial. One does not want to spend big dollars on a project if ownership is not established. Also, if done right, patents must be filed in all the major markets throughout the world and they all have different rules, so a patent law firm is absolutely required to sift through the morass of it all. Good patent attorneys command mid-to-high three-figure hourly fees, and the initial filing with the US Patent and Trademark Office or PTO is only the beginning of the process. This is why researchers and academia often have incomplete and weak intellectual property (IP) portfolios.

The road to an issued patent is long and rocky. Having been an inventor on 14 patents over the past several decades, the process is getting slower when it needs to be just the opposite. Before filing a patent it is necessary to do a prior art search to ascertain whether your invention is novel and does not infringe on previous patents. This is a long process, and the patent examiner will do one as well but the inventor should do their own to avoid filing something that is not novel. The most important part of a patent is the claims made at the end. This is where the attorney is really necessary because the wording must be precise and able to withstand potential infringement claims by smart lawyers and other inventors down the road.

Patent filing is followed by a series of comments or office actions from the examiner assigned to the patent application. Unfortunately it can take 2 or 3 years to receive the first office action. During that agonizing period an inventor is totally in the dark regarding what the examiner will say and again that may suffice for a widget but is totally unacceptable when a billion dollar venture is pending. When the office action does finally arrive, it will virtually always contain challenges to the inventor regarding the novelty of the invention, whether or not it has been reduced to practice and whether the claims of ownership are too broad with respect to the teachings within the body of the patent. The broadness of the claims and novelty of the invention always become major arguments between the examiner and the inventor because the broader the patent coverage the more difficult it is for future inventors to file competing patents. The examiner always argues for more narrow claims and again this is where a good attorney well versed in medical technology is critical. Novelty is also a major hurdle since it can be unclear whether the invention is obvious to those "schooled in the art" the patent deals with could have come up with the invention. The examiner often denies a patent for the obviousness reason, and such arguments can persist for years.

The first office action is meant to force the inventor to demonstrate all these criteria and begin the battle with the examiner. Multiple office actions are usually required to get a patent application approved, and the turnaround time on each office action generally takes a year or more since the examiner has a pile of applications reaching from floor to ceiling in his or her office. Furthermore, during this multiyear period it is not uncommon for the examiner to change due to turnover at the PTO, and this further slows down the process.

The first thing that needs to happen in the medical patent arena is to cut down the time it takes to get through the system. The wait time for review of a given patent with three office actions can run to 5 or more years, and this is simply not acceptable when cures are involved. Virtually all of the value in a medical start-up company resides in its patent portfolio. VCs, investment bankers, and private investors will not invest in companies that do not have a strong patent portfolio to protect its technology. This is especially true for medical R&D firms since the capital required to get a drug to market is so enormous and they have no sales income. The long timeframe also increases the cost and management involved in getting a good issued patent and makes it much harder on academia to build good IP positions on their technology.

So the number 1 action necessary on the legal reforms is to cut the review time on terminal disease-related patents. The first office action must be within 6 months of submission, and the inventor must respond within 30 days. This would greatly speed the pace of drug development by helping companies raise the capital for such activity. Also, I have found it useful to visit the patent reviewer assigned to a patent and walk that person through the science behind the invention and assist them with the existing prior art showing why the invention is novel and not obvious to those schooled in the art. I have found this approach to be very useful for a number of reasons. It establishes a relationship with the examiner who they appreciate and it makes their job a lot easier by taking some of the work load off their shoulders. It also avoids future office actions that could have easily been dealt with at such meetings.

OPENING THE FLOODGATES: PATENTING EXISTING SCIENCE

It is not enough to just make terminal disease patents easier and quicker to get allowed. The criteria for filing cure-related patents must also be expanded to include what is already in the public domain. After decades of failing to cure these diseases with novel new science, it is imperative that we look at existing, in the public domain, science, and the thousands of approved drugs that could be repurposed. This is the landscape that maverick entrepreneurs thrive in, and as we have seen in Chapter 3 this approach has produced impressive results. Imagine if this approach to cures was encouraged rather than marginalized. It is, however, almost impossible to do now because of existing patent law!

Of the thousands of drugs out there and the countless combinations thereof, only very few have been looked at as repurposed therapeutics. Part of the reason for this resides in the patent codes that offer weak protection for the necessary investment required to pursue the research involved. In addition to the old drug–new use drug repurposing opportunities, there is an enormous amount of published science that is not patentable. The reasons for this are several and they can all be corrected. First among them is the requirement that a patent be filed within 1 year of the technology being made public. This means that if a researcher publishes data either as a full paper or an abstract, they only have 1 year to file a patent covering the use or composition of that entity. This turns out to be a big problem for the inventor especially if they are an academic since their careers depend on rapid publication and not being scooped. It is publish or perish in academia, and patents are often a secondary priority and an expensive one to undertake. Patent filing requires specialized patent attorneys with both JD and PhD degrees, and investigators often do not have the financial support or the inclination to engage in the business of intellectual property. Most research grants also provide limited if any support for such activity as well as universities and research institutes. The net result is that enormous amounts of technology exist, which are totally useless all because of a simple requirement built into the patent law.

The simple solution that would bring this trove of science into the development pipeline for the purpose of developing a potential cure for terminal disease is to allow it to be patented by anyone with such a goal. This reform would reduce the reliance on new and unvetted science as the only substrate for development. Older published science has often been validated by other investigators and therefore is more likely to be successfully exploited. While it is possible to file utility or use patents currently, which enable utilization of existing science or compositions, they are not well defended in the market place and generally viewed by the investment community as being of limited value. It is also more difficult to get use patents allowed since they often conflict with the prior art infringement and novelty criteria imposed by the PTO. This could be easily remedied by reforming patent law codes to require more strict enforcement of such patents and relaxation of the prior art and novelty requirements. Again we confront here the same academic mentality in patent law as observed with the FDA in Chapter 9. The goal of medical patents should be the facilitation of developing cures and enabling those engaged in that activity, not imposing some academic construct of existing prior art or novelty, which only encumbers the end result.

BORN AGAIN PATENTS AND THE TERMINAL DISEASE EXCLUSIVITY ACT

Patents expire after 20 years or for failure to keep up with the payments to keep it active. This patent graveyard may also be due to the fact that the intended commercialization proved to be invalid or the inventor simply decided to terminate it. Whatever the reason for termination of these patents, they should still be able to be resurrected and refiled for a variation of the intended or a different use. There is no reason to remove this material from the attention of innovative entrepreneurs.

In addition to making it possible to patent existing science, a parallel approach to incentivizing the utilization of existing science that is not strongly patent protected would be to use the Waxman Hatch Orphan Drug Act as a model for terminal disease cures. This highly successful piece of legislation confers a 7-year market exclusivity period on drugs developed for diseases with under 200,000 cases per year in the United States. The rationale being that market exclusivity on a drug covering such a small market size would incentivize companies to develop drugs to treat such indications. This has led to numerous companies that simply focus on orphan drugs and the development of many drugs to treat orphan diseases [1,2]. Such activity would have never happened without this legislation, and a similar reform could be enacted for drugs or technologies that cure, or are disease modifying, for terminal disease. A disease-modifying drug is one that converts a terminal disease into a chronic treatable disease but falls short of being a complete cure.

This piece of legislation could be called the TDEA and could be approved by Congress quickly and offer an extended exclusivity period for any therapies that aim directly at such treatments. Perhaps a 15-year postmarketing exclusivity period would be appropriate. This would simplify the interaction with the patent office and reduce associated expenses and delays of filing and prosecuting patent applications in multiple territories. Such legislation would greatly increase to raw material for cure-related R&D, and numerous companies would reevaluate their business strategy since all the published technology in existence would now be available on a proprietary basis. The TDEA would draw much more attention to cures rather than me-too drugs. Its effect on the entire industry would be dramatic and greatly facilitate virtually all of the other reforms we have discussed.

While the TDEA would be a big help on its own, it also needs to be accompanied by the other reforms mentioned such as patenting published science, and resurrecting existing patents when the claims are for curing or modifying the disease course. Imposing legal barriers to drugs such as these makes no sense, and every possible means of facilitating their rapid development must be employed.

IMMUNITY AND INDEMNIFICATION

We have seen in Chapter 6 how easily drug companies can become targets for frivolous, fabricated, and extremely costly class action lawsuits and how difficult it is to defend such emotionally charged cases. Legal firms troll for clients and school them on how to act and what to say to win over juries. Companies are exposed to liability during the entire life of a drugs development and marketing, which not only increases the cost of the product but also delays its appearance in the market.

An especially troubling aspect of trying to get a drug approved is when a company reports a failed Phase III trial. Numerous law firms specializing in medical litigation often immediately pounce on the company for misleading its shareholders or not reporting various aspects of the drug to the shareholders. This results in a plummeting share price and extensive legal expenses for the company. The entire sham action occurs in spite of the numerous disclaimers every public company must make in every corporate document it issues from press releases to 10-K yearly and 10-Q quarterly reports. Such documents go on for pages enumerating the risks associated with drug development and the potential side effects that can be encountered. Also the product inserts of marketed drugs read like a potential death sentence is associated with the drug involved. In spite of this the company still is responsible for every hiccup the patient may encounter.

What the public fails to realize is that the FDA will literally crucify a company that does not report every aspect of a clinical trial and that the FDA has to approve the study design of any clinical trial. Any adverse event in a clinical trial must be reported to the FDA promptly, and the FDA can terminate a clinical trial immediately if it deems necessary. Yet the FDA has no liability and is rarely if ever sued. The company gets decimated. The solution is not to make the FDA liable but rather to reign in the lawsuits that ensue, which have made drug development into a defensive timid process rather than the aggressive venture it needs to be. The Vioxx, Avandia, silicone breast implant, and Tysabri cases previously discussed in Chapter 6 are examples of this legal system gone wild.

The grounds for filing such suits should be strictly defined. Medical class action law takes the term circumstantial evidence to new extremes. Clinical science is perhaps the most inexact of sciences and the lawyers know that. It is very difficult if not impossible to say with certainty that a drug caused a side effect. The fact that the incidence of a side effect is increased in the drug-treated group of a clinical trial, or a population if the drug is approved, does not demonstrate a cause–effect scenario in a given patient. Moreover even if such a cause–effect relationship were established, it must be kept in mind that the FDA is sitting on the shoulders of the sponsor evaluating and approving each move that is made. Unless clear intent and/or gross negligence on the part of the company or the FDA is demonstrated, both should be held harmless. Certainly this must be the case for terminal disease drugs where the risk–benefit calculus is unique.

In cases where the FDA does not recall the drug in question or where the company did not withhold any data from the agency, the company should be immune from prosecution. Just these simple and rational changes in the tort system would redirect billions of dollars to drug development and reduce healthcare costs dramatically. The far more important effect would be to encourage drug companies to attack the terminal diseases more aggressively and take more risks.

REIGNING IN THE MEDICAL TORT SYSTEM

Drug companies need more clarity on what they will be liable for. If they withhold information from the FDA or alter data, then they should be punished severely. They cannot, however, be expected to shoulder all the risks for a drug failing a clinical trial or a side effect it might show after it is on the market for 5 years. The punishment of having to withdraw a drug from the market if a safety issue arises is sufficient in most cases. These lawsuits must be reined in by making the loser pay for all expenses and putting caps on or abolishing settlements. The lawyers are always the winners in class action drug suits of which the great majority are resolved via a settlement. If the current medical tort system persists, drug companies will simply not take the risk to develop and market a first in class drug.

The public needs to realize that the pharmaceutical and biotechnology industry is among the most highly regulated in the world. Nothing escapes the eyes of the FDA—it is against the law, and companies that do it are at risk of being shut down for good. Unrestricted drug lawsuits do not serve the drug development that is needed to make quantum advances in treating and ultimately curing terminal diseases. FDA approval has to offer protection to the vicious and, in most cases, unjust retribution that lawyer-induced plaintiffs have sought from drug companies. There needs to be a limit to the liability a company has if we are to see the kind of risk taking that is necessary to generate breakthroughs. Until that happens, the tort system will be one of the major inhibiting factors to medical progress.

REFERENCES

[1] C. Helfand, Top 20 orphan drugs by 2018, Fierce Pharma. Retrieved from: http://www.fiercepharma.com/special-report/top-20-orphan-drugs-by-2018.
[2] E. Rensi, The orphan drug act as been a huge success, The Wall Street Journal (June 23, 2008). Retrieved from: www.wsj.com/articles/SB121417838559995535.

FURTHER READING

[1] When the peers all think the same(Letter to the Editor), The Wall Street Journal (August 6, 2014). Retrieved from: www.wsj.com/articles/when-the-peers-all-think-the-same-letters-to-the-editor-1407265925.

Chapter 11

The Public Will Make It Happen

Chapter Outline

Well, this is a big task but not as difficult as it may seem since the infrastructure is all there and just needs to be repurposed with the focus, strategy, and urgency that achieved the astounding breakthroughs in other technologies. The question now becomes how we make all this happen? Obviously this is a multifaceted problem, and fundamental reforms are required for all its components. A fault in just one will disable the result. The institutions we have built are entrenched, and the media has made them all "prestigious" without bothering to do a cost–benefit analysis or dare to ask the key question: What have you cured lately? We can fix this, but there is one more issue to discuss, you the public.

TAKING BACK CONTROL OF OUR LIVES WITH HEALTH SAVINGS ACCOUNT

The public needs to start the execution process with the ammunition provided in the preceding discussion. The scope of the issue requires an informed (hopefully this book) grass roots movement that raises this issue to a priority equal to that of national defense. In this case it is defending the 2 million victims who will succumb to terminal disease each year. It will be a fight with the vested interests singled out for reform since they are going to push back to protect the very attractive turf they have created for themselves. It must be made clear that this is not about more funding, but rather a fundamental change in how we do business. The best place for the public to start is to look at the healthcare system because that is where a serious cure effort will emerge and eventually end up. The wonder drug, if we are alive to see it, will be placed in the hands

A Roadmap for Curing Cancer, Alzheimer's, and Cardiovascular Disease.
http://dx.doi.org/10.1016/B978-0-12-812796-4.00011-8

of the Healthcare Management Organization (HMO) or insurance companies, and these payers will decide whether to pay for it. The public will have very little to say about it because the public is not a consumer of its own healthcare. It has surrendered that right to these third parties, and this is one reason why healthcare has become so expensive. We have spent a lot of time talking about how the R&D system impedes the generation of cures, but very little about the healthcare system that provides these drugs to the public.

With virtually every good and service the public pays for, it is in the driver's seat. If the service does not perform, one goes to another provider. Competition continually improves the product by giving the consumer what they want at a reasonable price. The drive to innovate is fueled by public demand, and it is unstoppable. The healthcare system deals with our lives, whether we live or die, yet we are far from being in the driver's seat, we are not even in the passenger's seat. We have absolutely nothing to say about the most important product we use.

How do we expect the pharmaceutical and biotech companies to respond to our needs if we never have any direct contact with them? We hire insurance companies or managed care groups to do all the talking for us, and they are primarily concerned about their bottom line and not us. They provide a service and lump us into groups that provide them with a profit and operate on the assumption that the consumer is ignorant of their own health. They have also totally usurped what used to be the physician's role and have relegated them to being soldiers that obey the guidelines imposed on them by their HMO or insurance company bosses. The physician–patient relationship we knew back in the 1950s is gone. No longer does he or she come into our home and get to know us, rather they stare at a screen peck on a keyboard and say next. We do not pay them directly, and they do not work for us anymore. This is almost funny given the "new" focus today on patient-centered care as if it was some new concept; well that is what it was back in the 1950s. It is true that the patient can switch payers, but the fact remains that the patient is still not the master of their health destiny, and physicians are increasingly salaried employees of HMOs. Single payer systems will make patient's rights even worse and further alienate physicians.

Imagine there were only one car company and that you could only buy that car from a third party. If that were the case, we would still be driving the Model T. This is exactly what is happening to cancer patients taking 50-year-old chemotherapy drugs or Alzheimer's patients taking Aricept. Without competition for the public's attention, progress does not happen. In the landscape of managed care and nationalized healthcare, which has evolved over the past several decades, it should be no surprise that terminal disease is still an issue. Rather than push on the healthcare system for cures our frustration has forced us to give to charities, which is not going to cure anything.

Now let us imagine that everyone had a healthcare savings account (HSA). This version of healthcare involves a catastrophic insurance plan with a healthy but affordable deductible to take care of the big issues such as hospital stays for major surgery but puts the patient in charge for everything else. Instead of paying a huge premium every month for healthcare you may rarely use, you deposit money into a savings account each year and draw on it as needed for noncatastrophic healthcare expenses. The money is deducted from your taxable income and unused funds accumulate over the years. This achieves a number of things that have a bearing on cures. First patients have a real incentive to take better care of themselves since they are paying as they go and not simply enriching the payer 24/7 for a service they may not be using 24/7. It also forces you to shop for healthcare, and the first question a provider would get is what is this going to cost? This will introduce a whole new aspect into the cost of healthcare, and costs will decrease dramatically. A cursory look at what the cost variation is for surgeries, and major diagnostic procedures will reveal enormous variation in cost indicating that some providers are much more efficient than others or perhaps less greedy.

With an HSA the physician–patient relationship is totally different. The physician is now on your side, free from interference by payers, and their rules about what is covered or capitated. This would have a trickle-up effect on the entire healthcare system. For example, if you wanted an MRI or other expensive tests, there would now be a patient spending their *own* money for that test. The cost of that test would become a major issue, and there would be all kinds of competition sprouting up. This is already happening with services such as life line screening that offers batteries of diagnostic tests that can be purchased on a fee for service basis without having to convince a payer that you want them. Having used these services, they are very well done, and the cost is a fraction of what is available through a third party payer organization. Multiply this by millions of patients and you would see the system adapt to that with lower prices, better service, and more centers providing that service. Currently the status quo is "do not do tests that are not absolutely necessary," well why should a bureaucrat tell you that and not your personal physician? The more tests the better when early detection is a big part of the battle as is the case with all three terminal diseases.

The bottom line with HSAs is that the patient now has skin in the game. For the first time in 60 years, there is a consumer in the picture paying attention to what is going on and not a middleman with ulterior motives. Patients become true partners with their physicians and that is patient-centered medicine in its purest form. Healthcare has unfortunately gone in the opposite direction, and with the advent of the Affordable Care Act (ACA) it has moved even further, putting everyone literally in one box and further removing the patient from the product. Every patient is unique in their response to drugs and treatments, so the one-size-fits-all approach to healthcare is idiotic. If the public does not get involved as a consumer with healthcare, it will be more difficult to change the R&D system.

KNOWLEDGE IS POWER

The public gets what it asks for. To know what it needs it must be informed and able to advocate for it. A major goal of this book is to convey information about the terminal diseases themselves and the R&D system that is trying to cure them. Power is equal parts caring and knowledge. When that happens, things get done. When it comes to healthcare and terminal diseases, the public needs to do a little research as it does for other products. After all, this is about living to 60 or to 96, not a trivial issue, and certainly more important than which washing machine or car you should buy. The average person has no idea what cancer or neurodegenerative disease is all about until they get it. They have no clue about the policies of the Food and Drug Administration (FDA) or the blatant lack of strategy in the fight to cure terminal disease. Those who do take the time to learn only do so when the dreaded diagnosis is on them, and it is too late to do anything about. They believe that medical research is doing everything it can and that the medical–industrial complex is the best in the world. If it was, we would have cures and not 50 different kinds of pain relievers. With instant access to information at virtually everyone's fingertips, there is no reason for the public to be naïve and dependent. The more we learn about the abysmal state of the medical research community, the drug industry, and the regulatory–legal systems, the madder we will get. Channeling that knowledge will give life to the revolution that this book advocates.

The average college student graduates with precious little knowledge of basic science. The student rebellion that led to the liberalization of university curricula in the 1960s and 1970s gave students much more latitude in custom-tailoring their education, and few choose science courses. As a consequence, students emerge from college with far less general knowledge than was the case in years past. They are trained to fit into certain career positions rather than to negotiate life. The lack of basic knowledge in biology is rampant in today's culture. Most of us do not have a clue about how the body works. If we did, we would realize just how stagnant the medical R&D arena has been and how far-fetched some of the newest technologies are when it comes to their potential application to cure diseases in our lifetimes. There is, however, hope for this condition.

One thing we do have today that was not available to previous generations is the instant accessibility of the Internet. Anyone with a computer and a connection to this virtual library can be redeemed, and rather quickly. The Internet makes learning instantaneous. The public must be informed and aware of the past history of medical research. If they were, they would realize just how much of a treadmill we have been on for far too many years. The slowness and lack of accomplishment would become apparent. Look up the most famous medical research institution, foundation, or university that you can find and enter the key word, cure, and you will find lots of stuff about science breakthroughs or how great their staff is but precious little about cures for cancer or Alzheimer's or anything for that matter. Many of these prestigious institutions have been in existence for 50 or more years, raised billions of dollars, and have absolutely nothing to show for it other than scientific data; much of which is unrelated to terminal disease and virtually all of which is not part of a coordinated strategy focused on curing anything.

For effective change the public must be informed and motivated. Information is readily available when one has a virtual library in their lap. The motivation part could not be clearer: either we rise up and change this sorry state of affairs, or we and our loved ones face falling victim to incurable disease. With the power of information and motivation, the public can now advocate, and public advocacy works.

THE AIDS STRATEGY

The AIDS situation deserves mention again as regards healthcare and the patient's role in making things happen. AIDS patients rose up as a group and would not take no for an answer. The congress listened; the National Institutes of Health (NIH) for a moment woke up to what its only mission should be, as did the FDA and bingo, something actually happened. The lawyers did not dare get involved and we saw an unprecedented decade of progress; AIDS is no longer a death sentence. The gay community exerted its rights as a consumer, and with the help of the media the lifesaving drugs quickly became a reality. This can happen with the terminal diseases, and the stakes are orders of magnitude higher given the number of lives involved. The government reforms required will only happen if the public demands it and makes it a requirement for their vote.

More health savings accounts (HSAs) would be a great first step toward reducing the cost of healthcare and transferring power from payers to the consumer. It would give new life to the patient–physician relationship time and give physicians the patient contact time to actually take part in the cure quest by experimenting with repurposing appropriate existing drugs on terminal patients in real-world settings. Practicing physicians played key roles in two cures, leprosy and peptic ulcers, a better record than the entire research establishment.

If what you have read in this book struck a cord and you are concerned with the specter of looming terminal disease, then write a letter to the White House and your representatives in the House and Senate demanding that the government make curing terminal disease a top priority and pass the necessary reforms of the FDA, NIH, and the legal system to make this happen. President Obama essentially did this with the delivery of healthcare when he took advantage of majorities in both the Congress and Senate to pass the ACA, so this can be done. It would in fact be much easier because the ACA required a massive cost felt both by the taxpayer and by those who do not qualify for subsidies and are now faced with increased insurance premiums. This is clearly not a partisan issue, and it is hard to imagine that the minority party would oppose such legislation. Let us make some noise, so our children and grandchildren will not have to live in fear as we have.

THE ROADMAP

In summary, here is what needs to happen to get serious about curing terminal disease.

Academia: Universities/Research Institutes

- Quality control must be a requirement for grant funding.
- Remove scientists from leadership/policy positions and replace with visionary and vetted managers.
- Establish accountability toward a cure as a requirement for funding.
- Abolish tenure.
- Greatly reduce indirect costs.
- Encourage mavericks and put thought leaders in perspective.

Academia: NIH

- Replace with the terminal disease administration (TDA) and make it the central hub for all cure research policy and strategy.
- Reduce institutes from 27 to 4. Install proven project managers in leadership positions to establish milestone driven metrics.
- Expand and fix a broken clinical center and bring in industry trained staff to run clinical trials.
- Upgrade and expand the required support facilities for drug development.
- Enhance focus on near-term applied research and adopt the appropriate perspective on unvetted and longer term strategies such as gene therapy and stem cells.
- Encourage mavericks.

Regulatory

- Reinvent regulatory strategy for terminal diseases.
- Remove terminal disease drug regulation from the FDA and incorporate it into the TDA.
- Greatly reduce clinical trial requirements (one Phase II), and cut the approval timeline/cost by 75% for terminal disease drugs.
- Restore patient's right to life by removing barriers to trying investigational drugs.
- Establish a Cabinet level position on curing terminal disease that would administer the TDA, Patent and Trademark Office (terminal disease related), and all issues relating to cures.

Patent Reform

- Open up all existing science to patenting including repurposing existing drugs and exploiting older currently unpatentable science.

- Pass the Terminal Drug Exclusivity Act, TDEA, with a 15-year market exclusivity term.
- Provide strict enforcement for all such intellectual property.

Drug Industry

- Incentivize the drug industry to focus on terminal disease cures rather than me-too drugs.
- Regulatory and patent reforms would go a long way in this regard.
- Tax incentives on cures.
- Extend patent term on cures.
- Less antagonist attitude toward drug companies.
- Medical tort reform, FDA approval indemnification.

Legal

- Reign in the medical tort system.
- Clearly define liability.
- Make losers pay for the winner's fees and lost revenue.

The Public

- The public must become a consumer of healthcare.
- Expand HSAs.
- Reestablish the patient–physician relationship and patient-centered medicine.
- Make terminal disease eradication a human right and lobby the government to make it happen.

Index

Printed in the United States
By Bookmasters

Journeys in Science

Journeys in Science
Inspiring the Next Generation

Fanuel Muindi
Department of Molecular and
Cellular Biology, Harvard University,
Cambridge, Massachusetts

Jessica W. Tsai
Boston Children's Hospital, Boston,
Massachusetts; Boston Medical Center,
Boston, Massachusetts

ACADEMIC PRESS
An imprint of Elsevier

Academic Press is an imprint of Elsevier
125 London Wall, London EC2Y 5AS, United Kingdom
525 B Street, Suite 1800, San Diego, CA 92101-4495, United States
50 Hampshire Street, 5th Floor, Cambridge, MA 02139, United States
The Boulevard, Langford Lane, Kidlington, Oxford OX5 1GB, United Kingdom

British Library Cataloguing-in-Publication Data
A catalogue record for this book is available from the British Library

Library of Congress Cataloging-in-Publication Data
A catalog record for this book is available from the Library of Congress

ISBN: 978-0-12-813090-2

For Information on all Academic Press publications
visit our website at https://www.elsevier.com/books-and-journals

Working together
to grow libraries in
developing countries

www.elsevier.com • www.bookaid.org

Publisher: André G. Wolff
Acquisition Editor: Mary Preap
Editorial Project Manager: Mary Preap
Production Project Manager: Poulouse Joseph
Cover Designer: Matthew Limbert

Typeset by MPS Limited, Chennai, India

CONTENTS

FOREWORD

Science is not only a disciple of reason but, also, one of romance and passion.
Stephen Hawking.

Scientists and storytellers seem to occupy quite different parts of the universe. We often think of opposites: truth versus fiction, nature versus art, counting versus recounting. Nevertheless, scientists and storytellers have a shared mission. Both help us to make sense of the world. Both help us to understand ourselves and our place in the world.

Our own stories also have great power. We tell ourselves stories in order to make sense of what happens to us. We log the consequential moments in the form of little stories. Our story moments add up to our understanding of ourselves and our journeys.

We know our stories, but we rarely tell them. We rarely share our accounts of the nudges, the turning points, the setbacks, and the successes.

What inspires you? What event set you on your path? Who asked you a question that kept you awake? Who offered you an opportunity to learn? Who encouraged your curiosity and creativity? What observation did you make that you couldn't explain—so you pushed for an answer? These are key moments of learning. They are the unplanned moments that shift the course of your life. Each of us is in the place we are today as a result of hundreds and hundreds of such moments.

As we confront the challenges faced by our species and our planet, we need all of the scientists and inventors that we can get. The power of human discovery and inventiveness is our hope.

It is persuasive to look at the numbers. It is clear that we are not tapping the widest swath of human talent. Women, scientists of color, and people from nonwestern nations are underrepresented. We haven't brought them into the fold. And we must.

This book encourages us all to put our muscle into being teachers and role models in order to engage every mind we can. Jessica and Fanuel, the authors of this book, argue that young people need inspiration and opportunities to become scientists. They implore us all to take on this mission. There are two things that each of us can do.

First, teach.

Teaching is a dicey occupation. The outcomes are uncertain and unpredictable. As a teacher, you try your best to reach your students. However, you can't be certain that you did. Different tactics seem to work for different people. On different days. And the outcome is difficult to assess. What does learning look like? Did the student learn what we want him/her to learn?

When I started teaching at the college level, I made my peace with this uncertainty by thinking of myself as farmer. I sow seeds. I cast my seeds far and wide. I try to pick the right days to plant: when the sun is out, when the soil is damp from recent rain, when the phase of the moon is auspicious. But sometimes, I sprinkle the seeds solely because it is the designated day of the week for sowing. Students look grumpy and disengaged. I am dispirited and distracted. The outlook for learning is grim.

But you never know. You really never, never know. What will a student remember? What will spark their curiosity? What will inspire? The lesson, then, is to keep at it. Doggedly. Keep teaching. Keep reaching out. Find the student who seems checked out. Maybe he/she is just shy. Maybe he/she has learned to turn that face to the world to avoid being teased. Don't give up.

Second, inspire.

What is a role model? My working definition is that a role model is a person who has accomplished something that you aspire to. Someone who shares your commitments. A person who lives their life according to values that you want to mimic. Sometimes role models are people you haven't met. A pioneering scientist. A courageous thinker. A social justice warrior. Sometimes role models are close to home: a parent or relative or a teacher. A role model can even be a fictional character. I confess that I REALLY admire Neville Longbottom's grandmother from the Harry Potter universe.

Role models do not need to be older than we are. Fanuel and Jessica are role models for me. They inspire me with their insight, their energy, and their accomplishments. As you read this book, you will learn about some of the things that they have done. Jessica has completed an MD and a PhD. My mind has always boggled at the idea of doing both of those degrees. Either is hard, but the combination seems impossible. Nevertheless, she persisted. Fanuel was born in Tanzania, studied in Great Britain and the United States. He started his own nonprofit while in his 20s. Each of them is working full time in challenging jobs. In their spare time, Fanuel and Jessica founded the STEM Education Advocacy group in 2014.

You can see why I count them as role models. They are making an impact. In several venues at the same time. In their professional work—healing children, advising graduate students. With their education advocacy. Encouraging children to become scientists. In their writing, reaching out and inspiring people that they haven't even met.

Each of us has the potential to be a model simply by striving for excellence in what we do. We embolden others when we fulfill our professional mission according to our personal values. We inspire when we do our best, and live with authenticity.

Of course we fall short of our ideals and aspirations. Our unguarded moments, when we are not at our best, may generate lessons that penetrate deeply. Life is tricky that way. Parents know that children are always watching and absorbing lessons from how we comport ourselves. So, it is important to be forthcoming about our shortcomings. Candor about our limitations is another necessary dimension of being a role model. When we describe the ups and downs of our journeys, we offer valuable lessons about perseverance, humanity, and humility. We are all works in progress. Just keep doing your best.

This book is written by two people who are both outstanding scientists and enthusiastic storytellers. It documents an extended conversation. As such, it is an invitation to jump in. You will feel—as I did—a desire to share your thoughts. To ask a question. To disagree. To tell a story.

The authors are reminding us to be teachers in all that we do. Despite the uncertainties and unpredictability, it is still imperative to put our full effort into being teachers and role models in order to engage every person we can. The possibility of unlocking the vast reservoir of human potential lies before us. There are students in every corner of the globe, and just around the corner from where we live.

Most importantly, this book invites you to tell your story. And then to listen to other people's stories. This is a way to inspire and develop the scientists and inventors we need for the future.

Chris M. Golde
Stanford University
BEAM, Stanford Career Education
Stanford, California, United States

ACKNOWLEDGMENTS

Jessica Tsai (JT)

You can't ever reach perfection, but you can believe in an asymptote toward which you are ceaselessly striving.

Paul Kalanithi,
When Breath Becomes Air [1]

I first have to thank Fanuel Muindi, my incredible coauthor and dear friend since our days as graduate students at Stanford University. I am grateful for our many conversations (and arguments) on neuroscience questions, scientific philosophy, the meaning of life, and the future of STEM education. I am so glad that we are both motivated by an inherent drive to improve society, founded in optimism and curiosity. I appreciate that we incessantly ask each other questions. At the core, Fanuel is a tremendous friend, and his persistently positive attitude is something that I always seek to emulate. He is an incredibly thoughtful human being, and every conversation we have leads to more provocative questions. We started this writing process in November 2015, and everything from start to finish has been a tremendous joy. I cannot thank him enough for his friendship.

To Moytrayee Guha who has been a critical component of so many of our heated conversations. I value immensely her unique perspectives in global health and admire her passion for child and maternal health in the developing world. Her travels have taken her directly to the doorstep of people in great need of basic health care. I am grateful for her willingness to always challenge the status quo and am so thankful for her compassionate friendship.

None of this would be possible without Elsevier who has guided us in publishing this book, allowing us to take this incredible journey. To Mary Preap, our editor, who has answered all of our questions, responded promptly to our emails, and maintained transparency throughout the entire publication process. We have learned so much about the world of publishing from her, and she has made this such an exciting experience.

To Nathan Vanderford for always believing and encouraging us, despite our seemingly crazy (and late night) ideas. He embodies a fantastic mentor and advisor, who consistently checks in regularly. He is not only willing but excited to engage in discussions on new ideas. The publication of this book would not have been possible without him.

To Chris Golde for taking the time to read our entire manuscript. I could not have imagined a better foreword to our book than the one that she has so meticulously and thoughtfully written. Thank you for being a part of this project and for pushing our readers to teach and to inspire. You are a role model for me and Fanuel.

To STEM Education Advocacy Group for being an open-minded sounding board for ideas. In particular, to Joseph Keller whose thoughts and ideas are a breath of fresh air.

To Dr. Claire McCarthy who has edited countless pieces of my writing (with many comments in red). Thank you for pushing me to continue writing purely for the joy of writing.

To Dr. Bob Vinci for always reading my writing and helping me to recognize that writing, basic science, and clinical medicine are far from mutually exclusive. Your confidence in me is often greater than my confidence in myself, and I am so grateful for your mentorship.

To Dr. Kate Michelson for so many things. From that first interview to many conversations ranging from coffee to implicit bias to writing to my future, your mentorship is truly exemplary. You have a remarkable capacity to listen and to ask precisely the right questions at the right time.

To Daena, Hoa, Jackie, Jamie G., Jamie L., Olivia, Osiris, Pam, Rebecca, and Renee for being incredible women. I appreciate and cherish your support and friendship more than you know.

To Imeh for your dear friendship and intellectual vivacity.

To Matt and Silas for always cheering me on (and sending me cute photos of sloths).

To Natasha for being the twin sister I never had. You are like family to me. Thank you for always encouraging me to work harder, ask more questions, and push my limits.

To Audrey for your optimism and many conversations about women in science.

To my dear coresidents in the Boston Combined Residency Program in Pediatrics, you all are incredible pediatricians. I am so lucky that I get to work (and dance) with you every day.

To HCH and TRC, two of my dearest mentors. Each of you has guided me on this journey through different parts of my scientific development. You both exemplify true mentorship, and the scientific community would be so lucky if there were only more people like you.

To Mom and Dad—two of the most remarkable people I know. None of this would be even remotely possible without all of the sacrifices you have made. So many of my scientific origin stories stem from my early childhood. Thank you for making the courageous move to the United States. Thank you for always letting me find my own path (even when you disagreed with it). You both epitomize hard work for me, and I strive every day to work as hard as you. Thank you for reminding me that passion is the key to everything, and that I will never be satisfied until I know that I have done my best. Mediocrity does not exist. You have also taught me that acknowledging the world's problems and even reading about them is far from enough. You have lent me tremendous empathy and a sense of responsibility for the suffering of others. Thank you for teaching me to be a doer. Thank you for showing me that it is more than okay to rock the boat. Positive change happens only with an optimistic attitude and a willingness to actually do something about the world's problems.

To Cynthia, my ever smarter younger sister, thank you for your kindness and warmth. My life was changed for the better the day you were born.

To all my patients, past, present, and future, thank you for teaching me more than I ever learned sitting in a lecture hall in medical school.

To young women—you can do and be whatever you want to be. You are brilliant, strong, and resilient—don't let anyone tell you otherwise.

To young children everywhere—you inspire me every day. Your sense of awe, your curiosity, your inquisitive thoughts—don't lose this.

I can't wait to see all the beautiful things you do for the world. Just because you are young, does not mean that you are weak or unimportant.

Godspeed.

Fanuel Muindi (FM)

To Jessica: You have been a wonderful friend to both myself and Moytrayee. Your friendship means a lot to us. We thank you for being there for us going all the way back to Stanford University. Thank you for the insightful conversations about life, science, health, death, and so much more! We have managed to capture a tiny fraction of those conversations between the three of us here in this book. Thank you for your continuous positive outlook and your drive as we continue to push forward on this journey of discovery. We look forward to many more discussions!

To Nathan Vanderford: This book would not have come alive without your help. Thank you for believing in us. Thank you for pushing us when we doubted ourselves. Thank you for your mentorship.

To Mary: As Jessica mentioned, thank you for your patience in dealing with all our questions. We were lucky to have you as our editor. As newbies to publishing, we have truly learned a lot about the world of publishing. Thank you for your guidance during this process.

To Chris Golde: I am glad we managed to touch base again all those years after the DARE program at Stanford. Thank you so much for reading our conversation and writing such an inspiring foreword!

To the STEM Education Advocacy Group: It has been fun working with you over the last 2 years or so. I truly believe we are creating something special with our group. Thank you for believing in the dream and being part of this journey. Remember, we are just getting started!

To Suraj, Ekta, Ajay, Shubha, Sanhita, Mark, and Naomi: Your friendship has meant a lot to myself and Moytrayee. I can't believe how fast the years have gone by. Thank you for standing with us throughout all these years. We are lucky to have such great friends support us the way you have supported us. This book is another small step and I appreciate all your encouragement.

To my parents: It all started with my passion for planes. Thank you for the wonderful gift that started this journey all those years ago. Thank you for seeing the potential that I had inside of me.

To Kamal and Minakshi: Where do I start? Thank you for being the best brother and sister-in-law I could have asked for. I am grateful for your continuous support to both myself and Moytrayee. You have had the pleasure of hearing a lot of my crazy ideas around science education. Thank you for patience! You have made the journey so much fun.

To Sid (my nephew): At the time of this writing, you were only 3 years old! It was wonderful to see your excitement and joy at watching all those space rocket launches on YouTube. Perhaps, one day we can go and watch one for real! It is my hope that you will eventually read this book in the not too far distant future. May it bring inspiration to you.

To my parents-in-law: Thank you for you infectious optimism and positivity that has helped a lot during the writing of this book. Your support has meant a lot in the last several years and I am deeply thankful for believing in myself and Moytrayee.

To Moytrayee: What a journey this has been. I do feel that you and I planted the initial seeds for this book from our many discussions surrounding education and global health going back 8 years or so. Time does really fly by! I have found these deep global discussions incredibly inspiring. I am glad you are the first person I tell all my ideas. I really do appreciate your honest feedback without which this book may not have come to life. You continue to push me to be better each day. I couldn't imagine a better partner on this journey. I thank you from the bottom of my heart for standing with me through it all. Love you lots.

To the reader: I hope this book inspires you to think back on your own journey. All it takes is one story to inspire a generation. Can your story be THAT story? We believe it can. This is just a start. May this book inspire you to tell your story.

Let us begin.

REFERENCE

[1] Kalanithi P. When Breath Becomes Air. Random House 2016.

Fanuel Muindi (F.M.) and Jessica Tsai (J.T.)

A journey of a thousand miles begins with a single step.

Lao Tzu.

J.T. — We can't get anywhere without taking that first step.

F.M. — And it is perhaps the most important and difficult one to take.

J.T. — Absolutely. Fear, shame, uncertainty—all of these feelings can cloud and impair our ability to initiate the first step. But the reward can be great once you get going on the journey.

F.M. — You said something very important. That first step is the beginning of a journey.

J.T. — That first step in the journey can feel so frightening. It can make us feel so alone.

F.M. — It doesn't have to be that way.

J.T. — So true. We really need to share the stories of our first steps with one another.

F.M. — Because it is easy to forget that all the great things that we see people achieve begin with that first step. Sharing stories about our journeys is absolutely important.

J.T. — So, what kind of stories are essential?

F.M. — Stories of how it all started.

J.T. — Stories of inspiration and discovery!

F.M. — Stories of the unknown.

J.T. — Stories of doubt and uncertainty.

F.M. — Stories of stumbling across important problems.

J.T. — Stories of struggle.

F.M. — Stories of overcoming those struggles.

J.T. — At times, stories of defeat.

F.M. — Yes, defeat.

J.T. — We never hear these stories.

F.M. – Not as often as we should. The stories mostly remain untold and many are ultimately forgotten.

J.T. – So, why are these stories so important?

F.M. – I have come to realize that stories have a lot of power. Stories have the power to inspire people, to push us to be better, to change how we see the world, to spark curiosity about the world that surrounds us, and so much more!

J.T. – Indeed. Stories also make us realize we are not alone in our experiences and struggles.

F.M. – Again, especially in science.

J.T. – But why do we not share these stories with each other en masse in science? There seems to be a lot of shame and embarrassment.

F.M. – I don't think it's shame necessarily. How about fear? The fear that we are the only ones going through such experiences. Many are afraid of highlighting their weaknesses or struggles, and so choose to only share their strengths.

J.T. – Well, I still think people are afraid because they are embarrassed about their mistakes. But in reality, it turns out that many of us have so much to share. And the only way to learn from one's mistakes is to share them with others and get feedback!

F.M. – We each have so much to contribute. Science needs this.

J.T. – Science depends on it.

F.M. – Yes, it certainly does. I think stories need to also capture important moments in one's journey in science. Was there a moment when you wanted to give up? Was there a moment your mentor told you something that helped you succeed? Was there a time you were challenged emotionally? Was there a moment when you felt alone only to find out you weren't? Was there a moment you felt like an impostor? Was there an "aha" moment in the lab? Was there a moment when you were inspired by the community around you?

J.T. – For me, "YES!" to all of these important questions. It's all about the individual moments that create a beautiful, complex journey.

F.M. – Exactly!

J.T. – Whose voices are we missing?

F.M. – Let's start with the young people. High schoolers for example. They have stories too.

J.T. – So do college students.

F.M. – Graduate students.

J.T. — Postdoctoral fellows.

F.M. — I mean, even administrators have stories in their service at academic institutions.

J.T. — Science journalists and bloggers as well.

F.M. — Yes!

J.T. — Aspiring scientists in the developing world.

F.M. — We definitely need to hear their stories.

J.T. — Doctors, dentists, nurses, and veterinarians.

F.M. — Why not! Even citizen scientists too!

J.T. — Professors also certainly have a lot to share about their journeys.

F.M. — This is a huge undertaking! But there needs to be the first step toward capturing all these stories. There needs to be that first conversation. Someone needs to take the first step.

J.T. — Our conversation in this book is our way of taking that first step to engage with the broader community. This is our first step to promoting an environment of openness and inspiration for young people to talk about science.

F.M. — Absolutely. So let us begin!

J.T. — How did the idea for this book even come about?

F.M. — Let's start with a short story!

J.T. — Do tell!

F.M. — The idea for this book was inspired from a conversation I stumbled upon between a monk and a philosopher. This discussion takes place in a book titled "The Monk and the Philosopher," where a father and son discuss a journey that transitions from a scientific to a spiritual quest [1]. The son in this story was a scientist who left science to explore Buddhism and eventually became a monk. I don't want to spoil their conversation so I highly recommend that people read the book. To me, this intimate discussion was a demonstration of how powerful stories can be. Here was a scientist who decided to take a path that you typically don't hear scientists take. What was even more impressive to me was that the authors of the book (the father and son) discussed this journey together throughout the entire book. I was hooked!

J.T. — I remember you telling me about this book. These are the discussions that people need to hear. Stories of convoluted journeys that lead you to new places in very unexpected ways.

F.M. — And to me, we need more conversations about these journeys!

J.T. − How about our many conversations? I love the conversations we've had in our book! We've had so many fantastic conversations on science.

F.M. − Exactly. It is these conversations that I think are the key. Like the monk and philosopher, we discuss our own journeys and discuss important topics that so many trainees in the pipeline ponder about. So to partially answer your first question, I think one of the core goals of this book is to bring about inspiration for science through stories.

J.T. − To think back to the humble beginnings in science.

F.M. − What we term "origin stories."

J.T. − The excitement and wonder of our beginnings!

F.M. − Yes!

J.T. − And stories that can inspire and strengthen communities.

F.M. − To pay it forward through mentorship.

J.T. − And create incredible new ideas.

F.M. − And also stories that cause us to pause and reflect.

J.T. − And stories that make us take action and empower young people not only in the United States, but all over the world. In addition to sharing our personal stories from STEM training and engaging people at all stages of training, one of our other goals is to emphasize that STEM education—beyond its importance for global economic growth, social advancement, and equity—is critical for examining how we could improve our thinking. For me, this has been a chance to think about what science should look like in the future and what we can do to make the scientific enterprise better. So as we share our scientific origin stories, how do we want people to read the book?

F.M. − With an open mind like the monk and the philosopher.

J.T. − Aha! I love it.

F.M. − The conversational approach we use here is something that I think the reader should keep in mind. These are conversations that I hope they will feel like they want to join. Perhaps what is most important here is that I hope the reader is somewhat inspired to start such conversations with those around them.

J.T. − Much like the conversations that we and our friends have constantly! Lots of friendly arguments and heated debates, but ultimately motivated by an interest in using science as a mechanism to make the world around us better. For me, it's crucial that young people feel compelled and encouraged to have these conversations and question everything.

F.M. – The idea isn't to push more people into becoming scientists. If it does then that's great but that's not the point at all. I see science as a vehicle.

J.T. – This is a key point, Fanuel. Science is a means to learning how to think critically. It teaches you to question everything. I love this quote from Pablo Picasso, "Computers are useless. They can only give you answers." What about the questions? There are so many lessons to learn through science by asking questions and not being afraid to doubt or be curious. I hope we convey that well through our many stories.

F.M. – Ultimately, I believe that by asking better and better questions, impactful change becomes possible. I think this is true not only in science, but across fields and life in general.

J.T. – Being willing and not afraid to change the status quo is the critical factor to affecting positive change. And you are absolutely right that this is true across all disciplines.

F.M. – It is this core foundation—asking better questions—that inspires me in science. I love hearing many scientific, and nonscientific, questions young people are asking. I love hearing stories of how some began their journeys in science by realizing the power and drive one gets from asking important questions.

J.T. – I love stories where despite the criticisms of others, people continue pushing, asking more questions, using their logic, doing the right experiments, and ultimately take us to the edge of our current knowledge. These stories make me realize that uncertainty is not to be feared. It should be embraced wholeheartedly.

F.M. – Well said. And the stories are not just ours.

J.T. – Absolutely. The stories we have collected from others are incredibly inspiring. I've enjoyed reading them because they describe so much of how people have been shaped by experiences. Seeing the world through the lens of another person is such an adventure and a treat.

F.M. – Those origin stories are quite telling.

J.T. – My other hope is that people can hear about our mistakes and flaws and learn from them!

F.M. – To pause and reflect as I mentioned earlier.

J.T. – Very true. I hope after reading these stories, people discuss, share, and think critically amongst themselves. This book should be a starting point and a launching pad for discussions amongst students, labs, and scientists.

F.M. – And even those not in science!

J.T. – Precisely. I can't emphasize enough that readers should use our conversation as a starting point. We should explain a bit about the chronology of the book, don't you think?

F.M. – Of course. There is some flexibility we have allowed for the reader.

J.T. – That's right. The book itself can be read in chronological order but it can also be read in snippets. Each chapter builds upon the last and covers different topics.

F.M. – So Chapter 1, Little Scientists, focuses on what initially sparked both of our interests in science ultimately concluding with how we decided to pursue graduate school. It explores the best strategies to pique the interest of young people and is an apt reminder of the optimism and wonder that STEM lends to youth.

J.T. – Chapter 2, The Local Perspective, identifies individual hurdles and issues that can arise amongst trainees at various stages of their STEM education. These include impostor syndrome, lack of self confidence, mental health concerns, conflicts with colleagues and advisors, mentorship, resilience, work-life balance, time management, and thinking beyond academia.

F.M. – Thinking globally is also important. We are now a global community. Chapter 3, The Global Perspective, hones in on STEM education at a global scale. We consider what STEM education could look like in the developing world and how an ideal training curriculum could be structured. This chapter focuses on concepts of community, data sharing, and collaboration.

J.T. – Chapter 4, Inspiring the Next Generation, encompasses the social responsibility we have as scientists and engineers. We discuss personal examples of paying it forward and highlight some interesting programs that mentor young people in STEM. This is where we share inspirational scientific origin stories we have collected from all sorts of people who just love science and engineering. We emphasize the importance of advocacy as part of any career as a scientist.

F.M. – And we use Chapter 5, STEM Education Advocacy Group: A Case Study, to discuss our own advocacy group as a case study to discuss the journey so far for the group. It is our hope that the story there will inspire others to think critically about how they can make a difference in their own communities.

J.T. – And we finish off with Chapter 6, The Future and Beyond, where we examine the future of STEM education and ask what we can look forward to with regard to data, collaboration, teaching, and learning. We explore what the future looks like and where we are headed.

And this is just the beginning, right?

F.M. – Yup. What we have done here is to have that first conversation. We are confident that it is not the last.

J.T. – Let us begin the journey!

REFERENCE

[1] Revel J, Ricard M. The monk and the philosopher: a father and son discuss the meaning of life. New York: Schocken Books, Inc.; 2000.

CHAPTER 1

Little Scientists

F.M. — I find it interesting that only a few people usually ask me about the "thing" that got me interested in science. So I want to go back to where it all started—Dar es Salaam, Tanzania. Back then you see, I really wanted to be a pilot. Seriously! Airplanes were a complete fascination for me. They still are for that matter. Watching planes take off and land at the airport with my brother and sister is something I fondly remember. I always wondered how in the heavens they managed to fly without dropping to the ground. I wondered what it would be like to fly. I think we also lived on a flight path because I would routinely see planes fly over our house. After finishing my homework—or sometimes before doing so—I found myself piloting the massive Boeing 747 or the beautiful 777 across the Atlantic while paying close attention to the altimeter, speed, position, fuel, and all the other gauges inside the cockpit. Of course, this was on the 1998 Microsoft's Flight Simulator, which routinely crashed our Pentium I computer. I thoroughly enjoyed plotting courses and guiding the planes from point A to point B without crashing. Anyway, I could go on forever. The simple thing is that it was a lot of fun. This was important. The game also allowed me to experiment and think about some of the math I was learning in the classroom. I think this is where I truly began to think seriously about the future and started answering the question of what I want to be when I grow up. It is fair to say my affair with science truly began in that virtual cockpit.

J.T. — It's funny—we have been friends since 2008 and discussed the ins-and-outs of science, but I never knew your origin story! I grew up surrounded by engineers in the heart of Silicon Valley and at home—my father is an electrical engineer and my mother is a software engineer. My father spent much of his career immersed in the semiconductor industry. When he came home from work, he would often bring me and my sister computer chip duds. These were chips that his engineering team had worked on in the clean lab but that were going to be thrown out. Back then, these were enormous chips! The size of a

Journeys in Science. DOI: http://dx.doi.org/10.1016/B978-0-12-813090-2.00001-4

dinner plate at times! We would snatch them out of his hands, and lie on the living room carpet fiddling with them, astounded by the sparkle of the silicon. It was the first time when I realized you could build something from nothing and it could do amazing things. I mean, you can build a bicycle, but it does one thing. You sit on it, pedal, and it takes you places. But a chip. It's so intricate, detailed, and beautiful. You build it, and it goes in a computer. It does anything you want it to do. I thought that was so cool.

My true love affair with science started in fourth grade. And I remember the actual day. It was earthworm dissection day. At that age, most of the kids in my class were incredibly grossed out by cutting anything open, let alone playing with the guts of a worm. I remember my assigned partner that day sat next to me, covered his eyes, and handed me the scissors. I couldn't take my eyes away. How was it possible that there was all of this stuff inside the body of an earthworm that could only be revealed by opening it up? And even more crazy to me was that somehow all of these slimy insides worked in concert to keep the worm alive. Now that was a novel concept for me.

Do you think your interest in planes and math as a child was motivated by specific people? I feel that I was nurtured to explore at home—my parents are engineers after all! And, I can identify specific people who inspired me as a child. When did you realize that doing science could be a job?

F.M. − I think I was one of those that was initially grossed out by cutting anything open. Although not scientists or engineers, I think my parents certainly played a major role in my pursuing science. You see, they did something interesting: they didn't get too much in the way of my tinkering.

I was recently watching a YouTube clip of Astrophysicist Neil deGrasse Tyson where he talks about what parents can do to get their kids interested in science. His one bit of advice: get out of their way!

Now, my parents didn't just let me do anything—in fact we had a lot of rules—but they allowed me to do the important things. This included spending a lot of time with my aunt and uncle who were university professors, getting computer access at home, playing video games, after school tutoring, and so forth. All these little things slowly pushed me toward science I think. Ultimately, they did send me abroad to the United Kingdom for high school where I further explored and confirmed my interest in biology. Why abroad you may ask? Well, I think my parents realized that I would have better opportunities to explore the world of science abroad. They were right of course. However, I do think it was a very hard decision for them to make both financially and emotionally. You see, I was only 13 when I left Tanzania.

This discussion gets me thinking a lot about the diversity of journeys into science. Some of us get into it earlier and others later. Some can pinpoint specific people that inspired them and the "a-ha" moment when it all made sense. Sadly, there are just as many, if not more, that don't have mentors or the early access to resources that you typically hear about as being crucial for getting into science. I really do think it is important to note that paths into science—the origin stories as you call them—are diverse and such paths are something that should be highlighted and celebrated more. Talking about these diverse journeys could be one of the ways to get more young people interested in science. It is not done as much as it should be.

J.T. – There are certainly times when I get frustrated—when experiments don't work, when a paper gets rejected for the third time, when one of my patients is really sick—that I reflect on those very early experiences. I agree with you completely that this is not done as much as it should be. I was recently observing a second-grade classroom. The students were learning how to identify triangles, quadrilaterals, pentagons, and hexagons based on the number of sides. I don't even remember learning that! We take these things for granted every day. I remember the first time I learned that the heart actually pumps blood to the entire body. What! How! Incredible!

My parents were always very hands off. My father had one mantra that he would always say to me and my sister.

"Attitude is the meaning of everything. If you don't have it, nothing is important."

I remember we used to mimic him saying this in a sing-song way as kids often do. But, my parents were satisfied if we were satisfied if that makes sense. My father, in particular, instilled in me this sense of knowing when I tried my best. No matter the outcome, as long as I felt that I tried my best, that was all that really mattered in the end. And, I think I took that to mean that I could pretty much do anything.

I can't imagine leaving home at the age of 13, especially since I didn't leave the Bay Area until the age of 28. And when I left, I only moved across the country, not to a different country! How did leaving home shape your way of thinking and how has it impacted you as a scientist?

F.M. − I think leaving changed the way I thought quite drastically. I came out of my shell. I was rather shy and leaving home was the jolt I needed to break free from the shyness. As the years passed, I gained a new level of confidence that I never thought I could have. Being so far away from my comfort zone forced me to grow in new ways. I certainly did feel like an outsider in the beginning for all the obvious reasons. But this changed as I grew more comfortable with the people, language, food, and, *um*, the British weather. It wasn't easy but I managed. To some of my friends, I was one of the "smart" kids—at least I thought they thought I was. The truth is that I actually found it rather challenging. I knew that if I was going to succeed, I needed to put in the extra work. As such, I spent a considerable number of additional hours trying to understand the concepts. Like you, I tried my best. In the end, it is the only thing you can do.

However, I did not feel that I could do pretty much anything. Did you really feel like this?

J.T. − I had some pretty funny ideas of what I would be when I was growing up. And these ideas evolved over time, of course. I went through a phase where I wanted to be a WNBA basketball player. I was always taller than the boys in elementary school (my late grandfather was very tall) and played a ton of basketball. I took dance lessons from an early age until I graduated from high school, and at one point even considered being a tap dancer or a hip-hop choreographer. When I say that I could pretty much do anything, I think what I really mean is that I could try anything. I wasn't nec-essarily good at everything! My elementary school put on these show-stopping musicals every year through the choir. And each musical had a few students who had leading roles. They acted, they sang solos, they got to wear special costumes, while the rest of the kids sang in the back. I remember trying out for a solo, and I was terrible! My mother knew I was terrible! But, she let me try out anyways. She let me have that experience. She let me put myself out there. And she did not dissuade me from failure.

> So, when I reflect more about it—feeling like you can do anything really means acceptance of failure and a recognition that failure is not only okay but should be encouraged.

I never realized that, of course, as a child. Trust me, I balled my eyes out when I didn't get a solo part in that musical!

F.M. – You bring up a really good point here. Failure. I will be honest. I was terrified of failing especially given that failure meant—at least I thought—a trip back home for me. I couldn't imagine having to explain myself to my parents. But of course, the education system is fixated on students avoiding failure.

> Failure is considered bad. Of course, failure is how science works. We learn through mistakes and failure. We learn a whole lot more through failure.

This is something I never understood when I was younger and I am only getting it now.

J.T. – I only started appreciating failure in graduate school when failure was the status quo. It's interesting because in school, we are not rewarded for taking risks and being bold. Rather, we are rewarded for doing what we're told to do. I remember doing labs in science class in elementary school, and there was always a "right answer." One could argue that you have to learn the scientific method first—you know, generate a hypothesis, come up with your methods, execute, analyze your data, and present your data in some graphical or representative form.

> I've thought about this a lot, and I think if you give kids the opportunity to fail, the most important thing is that you help them learn from their failure. You catch them when they fall. That's the only way that they will pick up the pieces, cut their losses, and keep trying!

Do you remember your first failure?

F.M. – Interestingly, I don't. I think it was probably so traumatic that my brain blocked it. I suppose it probably involved getting a horrible grade on some test at some point.

J.T. –Yeah, I don't remember mine either!

F.M. – I think very few kids remember their first failures. It's the trauma!

J.T. — Did you participate in science fairs? I think they are well set up for kids to fail in constructive ways. Believe it or not, I actually won my fifth grade science fair although it was a fairly simple project. I used my mother's hair as a proxy for humidity. It was basically a human hair barometer that I left on our back porch. And every morning, I would run out there and take the measurement before going to school. A science fair project is certainly not the same as graduate school! However, what the science fair taught me was data collection. And diligence. It was important to go out there at the same time every day (that included weekends ugh!) and to be very exact in my measurements. I kept these logged very neatly on a sheet of paper and made some funny-looking Excel graphs to report my data. I have this picture somewhere in my parents' house of me standing in front of my project with all of my friends, everyone with the cheesiest grins on their faces. I remember being really surprised that my friends thought winning the science fair was actually cool.

F.M. — I never actually participated in large science fairs or competitions. I kind of wish I did but I think this is something that I was not exposed to in Tanzania or the United Kingdom back when I was young. The interesting thing is that there are so many of them nowadays. The Google science fair comes to mind as one of the good ones. The question then is which students end up participating in these fairs?

J.T. — That's a great question. The science fair I participated in was quite small, and it was only within my school. The purpose was meant to be fun and engaging, to work on a longer term project, and everyone was applauded for just participating. I had never worked on something for a prolonged period of time like that. I always wonder about those big science fairs that you mention. It seems that children who have parents in academia or are connected to faculty members at research institutions would have a significant advantage. Are we just selecting for privilege [1]? I'm not sure...

F.M. — I really think that all it takes is being aware of such opportunities. I would slightly disagree in that I don't think parents in academia or those who are connected to faculty members at research institutions have a significant advantage. Yes, having parents in the sciences helps a lot but I think school plays just as big a role—if not bigger—in exciting students about science and getting them to participate in events such as science fairs. Of course, I am talking about the teachers. One program that I am really excited about is the New York-based 100Kin10 nonprofit which unites the nation's top

academic institutions, nonprofits, foundations, companies, and government agencies to train 100,000 new STEM teachers by 2021 [2]. I
really think teachers play a rather important role in getting kids excited
about science. Teachers can encourage their students to participate in
science fairs if they are not already doing so.

J.T. – Agreed completely! Teachers are the true heroes. Every child
has incredible untapped potential that can certainly be brought out by
parents or peers. But in the absence of that, teachers are often the ones
who truly spark curiosity, encouraging students to ask questions and
take risks. A friend once described to me that a teacher should
be analogized to being a Sherpa or a guide. They ought to let you take
your own winding path but should most certainly nudge and push you
so that you learn valuable lessons without completely falling off the
mountain.

F.M. – I think initiatives such as the 100kin10 highlight
America's realization that tomorrow's workforce requires a strong
foundation in science, technology, engineering, and math. You may
be surprised to learn that there are some 6000 or so STEM programs
in the United States. This includes stand-alone programs like
100Kin10 as well as multiple programs administered by larger agencies, governmental entities, and nonprofit organizations. A really
cool website I found called STEMconnector provides a comprehensive directory and analysis of these programs across the United States
[3]. I think the gateways to science and beyond are plentiful these
days—at least in the United States. But there are still issues. For
example, in 2012, the number of high school freshmen expressing
interest in science was estimated to be around 1 million students
(something like 28%). Interestingly a little more than half of those
students were expected to lose interest by their final year of high
school [4]. Sadly I do believe that interest in STEM is even lower
when it comes to young women and underrepresented minorities in
the United States. It is something we hear a lot about. I keep mentioning the United States in particular because there is more openly
available data in comparison to many other countries. I often wonder
how interest in K-12 science differs across the world. I am fairly certain there is a survey out there for this.

J.T. – That's definitely true. I do think environment and opportunity are key. One of my favorite studies is the Hole-in-the-Wall study
where computers were set up on the street in different slums in India
[5]. Children with no experience and no exposure to computers were

actually able to learn to use the computers. The concept, referred to as "minimally invasive education," is an incredibly powerful one!

F.M. — It's amazing how such simple ideas can have such a powerful impact! It doesn't take much to get that initial spark at a low cost.

J.T. — Some of the most impactful ideas are the least technologically savvy!

F.M. — Very true. I have been doing a survey of STEM programs in Africa and it's amazing how diverse they are [6]. There is a definite focus on getting more girls interested in science which I think is amazing. The number of programs is nowhere near the 6000 + STEM programs in the United States alone, but I suspect this will change in the near future. At least, I hope it will. Anyway this is something we can explore further later. I really do think that getting the whole world excited about science can bring a lot of good.

J.T. — This survey sounds very promising! Definitely a huge undertaking but a worthwhile one, no doubt. I'm curious what STEM programs look like in other parts of the world. As a global community, we have a lot to learn from one another in terms of clever and savvy ways to create STEM opportunities for young people. We have so much knowledge to share with each other. I envision a global STEM summit, something like a meeting of the minds where people gather to discuss and brainstorm ideas!

F.M. — I like it!

J.T. — Clearly, STEM exposure as a kid is critical.

F.M. — Very critical.

J.T. — What were your STEM experiences in high school? I would describe mine as fairly prescribed and regimented.

F.M. — I found myself very interested in the sciences and math. I was particularly passionate about biology. I was good at it. I wasn't the best but that wasn't important. What was important I think was that my teachers kept encouraging me. Yes, it was a little prescribed and regimented at times, but I found my way through.

J.T. — Encouragement is essential. I've always felt it's important to have people in my life who push me through and believe in me even when my own confidence in myself is shaky. I had great teachers who made the subject matters interesting and fun, but I definitely spent a lot of time memorizing. Did you take Advanced Placement (AP) classes in the United Kingdom?

F.M. — The UK system was running the A-Level system at the time. Basically similar to the AP classes. At Ashville College (my high school), I found myself loading up on the math, biology, physics, and chemistry courses for the A levels. I didn't shy away from these courses for some reason. You always hear about students shying away from such courses. I would say that students should not be so afraid. Yes, it will be hard and you may even get a bad grade here and there. But you have to make the effort. Dr. Benjamin E. Mays (sixth President of Morehouse College in Atlanta) said that "not failure, but low aim is sin." Again, I think my teachers were crucial here. I didn't realize it then, but their continued support was crucial. Of course, it certainly helped that I found the content very interesting!

J.T. − That's awesome. My high school had many different AP courses, and I too loaded up on these because I found them to be more challenging and stimulating. However, I think there was quite a bit of focus on AP test scores. There was a lot of pressure to "perform." By the end of high school, I was really in a place where I was ready to advance my learning beyond memorization to critical, complex thinking, and problem solving.

F.M. − Students need opportunities to go beyond memorization. Sure, one needs to memorize things. Yes, they also need to problem solve. We always hear these things. But we can even push further. I do believe there can be structured exercises whereby students can find problems. Yes, problems. I mean, why not?

J.T. − Yes! I love the idea of finding more problems. Expand more, what do you mean by that?

F.M. − For me, the question is: How do we teach young people to ask important questions? How does one generate new knowledge from old knowledge? How does one identify new questions whose answers will provide new ideas and modes of thinking? We should be encouraging these at all times.

J.T. − I always thought it was odd that school was so focused on finding THE answer. When in reality, persistent questioning should lead you to ever more questions! I hope one day this will be incorporated into early education.

F.M. − I think it is incorporated to some extent but the issue is consistency. But what about the next steps? How was your STEM experience molded and shaped in college?

J.T. − Leading up to college, my STEM experience was purely from classes in school and interactions with teachers who really captured my interest early on. I was totally clueless about research, and honestly was not quite sure what I was going to do after college. I had always thought the brain was very cool (and enigmatic) but also loved marine biology and the outdoors, having spent a lot of time in Monterey and Santa Cruz as a child. All I knew was that I really liked biology. In the spring of my freshman year, on a complete whim, I applied to an undergraduate summer research program through Stanford's Department of Biology. I magically was accepted, and the die was cast. Looking back, I seriously cannot believe how fortuitous that summer was. I was incredibly eager and an entirely blank slate. Reflecting on it now I was really poised to soak up as much information and knowledge that I could. But, I seriously did not know anything—I was absolutely lost. It's funny thinking back now that I had no idea how to make an agarose gel, I barely knew what PCR was, and somehow I found myself in this crazy new world scribbling in a lab notebook. I wonder how many colleges and universities have these more structured summer research internships. I was grateful because I got to work in a lab plus there was a stipend to ameliorate some costs, and on top of it all, the program had some seminars incorporated into it throughout the summer. What was your first research experience?

F.M. – So, I actually wanted to be a doctor—that is a medical doctor. I didn't know a thing about research but it was something that was encouraged and so I thought, why not! I applied to a few places around the country in my sophomore year of college. I hear high school students are getting summer research experiences these days! I am sure some high schoolers are already authors on some publications too! Anyway, as an international student, getting a summer research internship was a little tricky. This is something a lot of international students continue to struggle with today. Many of the programs are federally funded thus restricting participation to U.S. citizens and permanent residents. Anyway, Georgia Tech accepted me somehow and I found myself in the School of Chemistry and Biochemistry. Like you, I had no clue what I was doing in the lab but I had a great mentor who was a graduate student. She made it a little less intimidating for me.

J.T. – Wait what?! I didn't know that you wanted to be a doctor! Gosh, I've also heard of high school students doing these summer research experiences. I sure wish I had done that! My summer in the lab honestly changed my life. I was totally captivated, constantly mesmerized, and just fascinated. Everything was so unstructured—it seemed like we were literally creating something from nothing. One of the things I never anticipated was that the lab would become like a family to me. Over the summer, we would BBQ together on the weekends, always celebrated publications and birthdays, and often went to grab a bite to eat after a long day in the lab doing experiments. Up until this point, I had only been exposed to didactic teaching where someone would lecture at me or read off of PowerPoint slides.

F.M. − To be honest, I was 70% terrified and 30% excited about the research and program overall.

J.T. − I was 100% terrified!

F.M. − What were you most freaked out about?

J.T. − I think the novelty of everything. I walked in that first day, and everything was alien to me. What was your biggest fear?

F.M. − Don't get me wrong, I was a good student and all but, for some reason, I still thought I didn't have what it took to be there. My impostor syndrome at this point was rather persistent. So, my fear—irrational I admit—was that they would find out I wasn't as good as they thought I was. Did you face the impostor syndrome during your early research experiences?

J.T. − Absolutely. I remember there were other students in the lab who were much smarter and more knowledgeable than me. I was so intimidated. I was waiting for the day when they would tell me they accidentally accepted me into the summer internship, and that I should just go home.

F.M. − I think people underestimate the paralyzing power of the impostor syndrome. Well, we now know it exists across ALL levels all the way up to CEOs.

J.T. − Oh definitely. Isn't it funny how feeling like an impostor resonates with so many of us, but we often still feel alone in our fear? We really do need to talk about this more openly.

F.M. − Very true and it doesn't have to be the case that very few people are talking about the impostor syndrome. I think more students need to understand that they are not alone.

Many of us feel like an impostor but the key is to not allow this feeling to paralyze you!

We should talk about it more often.

J.T. − What do you think the impostor syndrome is based upon? Is it fear of the unknown, fear of failure, some combination, or something else entirely?

F.M. — It is a mixture. I do feel that I had a committee in my mind at one point that was louder than my own true voice.

J.T. — Ah yes, I remember reading your piece that you shared with me. Sometimes we put the most pressure on ourselves, rather than outside forces putting pressure on us. What is the best way to grapple with this? I think undergraduates often experience this for the first time in their lives. There used to be a saying at Stanford called the "duck syndrome," where everyone seemed to be swimming along peacefully above water but kicking hard to try to stay afloat beneath the water.

F.M. — I do think we all have "committees" in our heads with some being louder than others. I still do believe that we—especially those early in their training—should learn to tell the negative committees in their heads to shut up from time to time.

J.T. — Yes, sometimes we have to tell those voices to pipe down!

F.M. — Like seriously. But how do we do that?

J.T. — I think the key is outstanding mentorship and peer support. And not being silent about the committees in our heads. It's so important to talk about these issues aloud.

F.M. — Yes! There is so much negativity out there that a little bit of honest encouragement and constructive criticism goes a long way.

J.T. — Agreed, my best learning moments during that first summer were really when the senior research scientist I was working with would sit down and explain a new concept or experiment to me for the first time. Really investing in my learning helped me to feel a part of this scientific community. And ultimately I think feeling insecure is okay and normal, but we have to be secure with our insecurities.

So why didn't you go to medical school? It's not too late you know!

F.M. — By the end of my junior year, I was fairly certain that I wanted to apply to the joint physician—scientist training programs out there. One, I wouldn't have to choose between my two interests. And, two, medical school would have been free! Sounded like a good deal to me. So, I applied to something like 4—5 programs and pretty much no one took me.

J.T. — Their loss!

F.M. — They did me a favor actually. It was the universe's way of telling me that there was something else for me. It was a little silly applying to so few programs anyway. I think Alexander Graham Bell said it best: "When one door closes, another opens." Well, something to that effect. The moral of the story here is to remain flexible. The situation was different for you though right?

J.T. — Isn't it amazing how things seem to work out for a reason? We often don't have the foresight to see that far into the future! I'm always amazed when you ask scientists how they got to where they are now. It was rarely intentional.

F.M. — For some, their journeys are probably linear but for many it certainly isn't.

J.T. — It was different for me—I really couldn't decide between the two, and ended up applying to joint MD-PhD programs. I was pretty fortunate because I had a physician—scientist mentor who guided me through the process, and gave me a sense of what life as a physician—scientist could look like.

F.M. — Again, mentorship.

J.T. — Precisely.

F.M. — Something I forgot to mention was that in the midst of applying to graduate programs, I got an opportunity to visit Stanford to learn more about their graduate programs. Now, December was rapidly approaching mind you. Most deadlines are around that time. Anyway, I met a lot of cool people during my visit at Stanford and of course I knew I had to apply. I took the GRE and sent my application in and hoped I would be accepted. You see, another thing that attracted me to Stanford was their Master of Science in Medicine Program (also known as the MoM program) which I found rather interesting [7]. I was still attached to medicine, you see. Long story short, I was accepted to Stanford and thus the formal journey into science began. Looking back, things were a little rushed and last minute.

J.T. — Wow! What a fortuitous series of events. The MoM program is truly an attractive one. What do you mean things were rushed? That you wished you had taken time off?

F.M. — To be honest, I really didn't think things through in great detail with respect to applying to programs. In a way, I was at the right place and it was the right time for me. Perhaps, more importantly, I was prepared for the opportunity. I think what I am trying to say is that I could have planned things even better and still left room for the unanticipated events. Again, my advice to those in the pipeline is to spend some time thinking about what you want to do in great detail. You don't have to figure it all out in one go. However, some planning is required. In the midst of all this, I strongly recommend you find good mentors who will be able to help you strategize and plan accordingly.

J.T. – The perfect storm. I know what you mean though. I try to tell undergraduates who ask for advice that graduate school is not a filler. You need the drive from within to do it, otherwise it won't sustain you in those late nights or the majority of time when experiments just don't seem to be working. I completely agree with you about good mentors. It's funny, people considering graduate school often worry about the topic they will study. I always say it actually matters very little. Rather, you should have an excellent mentor who will teach you, through whatever topic you study, to think critically and ask the right questions.

F.M. – The topic matters too. I think one of the issues though is that some students place all their emphasis there. One should spend time thinking about mentorship as well.

J.T. – What are some considerations to look for in a mentor and in a lab? How did you choose your PhD lab?

F.M. – The topic of research was actually what really interested me. I came into the PhD program with a strong bias toward working with a specific principal investigator who ultimately served as my PhD advisor. After visiting the lab, I became more sure that this was where I wanted to be for the next several years of my life. The lab was relatively small, and his mentorship style—a little bit hands off—was what I was looking for. It worked for me so I joined. What about you?

J.T. – I had thought about this in a fairly systematic way.

Mentor: I wanted somehow who was around some of the time, but not all the time. A mentor who would be available to me whenever I needed guidance but who wasn't constantly badgering me for my data. It was also important to me that this person cared about my scientific interests but also was interested in life outside of the lab and actually cared about my well-being.

Model system: I was very interested in broadening my skill sets. I had done quite a bit of work using mice to study sleep neurobiology, so I wanted to use a different model organism such as *Drosophila melanogaster* or *Caenorhabditis elegans* to study neuroscience in an extremely genetically tractable manner.

People in the lab: I wanted a lab that had a track record of having previous PhD students, including MD-PhD students. That was very important to me. I wanted a mixture of graduate students, postdoctoral fellows, and undergraduate students. I felt that I could learn a lot from postdoctoral fellows who had come from different PhD backgrounds, but I also really valued the camaraderie of having other PhD students. Having undergraduates was important to me because I wanted to have the opportunity to work and mentor students as well.

Size: I was aiming for a lab that was 10−20 people in size. I didn't want something too small but I didn't think that a really large lab on the order of 50 people suited my personality.

F.M. − I think you hit all the key points there.

J.T. − You know something that I had totally not thought about later was tenure and financial security.

F.M. − Neither did I.

J.T. − Six months after I joined the lab, my advisor got tenure. And I thought to myself—"I hadn't even thought about what would have happened if he didn't!" Financial security was something I took for granted as well. Being able to buy whatever reagents and try random experiments is really a luxury, and the reality is that it requires grant money.

Were there any other considerations that you only thought of later?

F.M. − Let's see. Average time to graduation and what I would be doing after graduate school. I guess it sort of makes sense to have such thoughts. However, I think all of us—my PhD cohort, that is—were just excited to have been accepted to graduate school and those things didn't matter that much at first.

J.T. − Wow, I definitely had not thought of that! Totally, we were all just excited to be there! I will say I did talk to graduate students in the lab who were finishing and on their way out. I think they were invaluable in providing the inside scoop and solid advice.

F.M. − I totally agree. I think students should just do their homework before making a decision. Don't assume anything.

J.T. − Doing your due diligence is key. Hopefully the working relationship you have with your PhD mentor and lab will be a lifelong one! Ultimately, though, I think I ran into issues with asking too many people for advice when choosing programs. Everyone had their own very strong opinion, and it's important to be able to filter that advice and take ownership of your decision!

F.M. − Yup, my PhD advisor actually flew across the country to attend my wedding! Mind you this was after I had graduated! It meant a lot to me. Anyway, I also think that due diligence must be done when it comes to selecting graduate programs. Of course, this applies if you are fortunate enough to have multiple offers. One should ask graduate programs the hard questions: How is the program's mission aligned with its values? What are the training outcomes? Where are the PhD graduates? How long did it take them to graduate? What percentage is able to do industry internships during the PhD program?

Any others?

J.T. − Finding out where graduates are now is extremely important. For me, I also wanted to know the track record for MD-PhD graduates. Where did they go for residency? What was the average total length of the program (it's a long training path)? Did anyone go straight into a postdoctoral fellowship? Are alumni now practicing medicine, running research labs, or doing both? And, if they are doing both, what percentage are they split between clinical and research duties? These are all important considerations.

F.M. − What do you think about having some of the answers to these questions available on all program websites?

J.T. − Absolutely. I think transparency is key. It would be really cool to have that information in one place, almost like a database.

Entering graduate school is a decision that is not to be taken lightly, and I think applicants and students deserve to make the most informed decision they can. I always appreciated in interviews when I met with people who were more interested that I make the right decision for myself rather than necessarily recruiting me to their program even though it might not be the best fit.

F.M. − Making the right decision for yourself is very important given that you are likely to spend 5−6 years of your life in the program!

J.T. − Did you have a gut feeling as well? Sometimes, you have all the right data in front of you, and on top of that, you just seem to know. I think it is important to listen to that feeling.

F.M. − I think I kind of did. Of course, my options were rather limited but I didn't complain. For some, they may only have one option at the end of the day. They may be forced to settle.

J.T. − It's true, in that case, information is of the utmost importance.

F.M. − Also, things can be complicated for those who do not get accepted anywhere on their first try.

J.T. − That is a really good point. And it is okay to try again, but I think it's really important to get feedback. Dig deep within and decide if this is what you really want. And if it is, then find out how you can strengthen your application.

F.M. – So, if you could have a do over, what would you do differently when it came to applying to graduate school?

J.T. – Hmm, I think I would have done more background research. I honestly picked programs based on name and did not really look much into the details. It was pretty random, in retrospect, which was really not strategic at all! I do distinctly remember though asking for MD-PhD program residency match lists when I was trying to make a decision. For me, that was a good measure of how their graduates had done based on what programs they matched into.

F.M. – As I mentioned earlier, I think it is crucial to do your homework and not leave things to the last minute. Do not procrastinate! As we talked about earlier, gather as much intel as possible on graduate programs! How long do their students take to finish? Where are they? Does the department even have such data? I could go on. Anyway, create a master file somewhere.

J.T. – Definitely, definitely.

F.M. – I think it is important to also remember that everyone has their own story in science. People forget that sometimes. Some of us are international. Some don't have the strongest science backgrounds. Some come from tough economic backgrounds. Some are first generation PhD students. We really are a diverse bunch just like any other field. However, we are united by our passion for science and our desire to use science to advance humanity.

J.T. – And that diversity is so incredibly valuable. Bringing all of those backgrounds together in the lab allows for amazing collaboration and innovation. It's like a melting pot of ideas. There is so much we can learn from each other if we embrace our differences. And you're right. At the end of the day, we all have a passion for science. For me, it's so important to remember my "origin story." Reflecting on the first time I was totally in awe and astounded by science is so inspiring. I always think about that sense of discovery and wonder when I'm working hard or when things aren't going quite right. While you loved airplanes and math, I was busy dissecting earthworms!

F.M. – It is easy to forget one's origin in science but it is crucial that such inspirational stories are shared with those coming behind us. Of course, everything we have talked about so far is leading to the one period every scientist remembers very well.

J.T. – GRADUATE SCHOOL. The agony. The ecstasy. Haha!

REFERENCES

[1] Flanagan J. Science fairs: rewarding talent or privilege? PLOS Blogs 2013; http://blogs.plos.org/scied/2013/04/15/science-fairs-rewarding-talent-or-privilege/ [accessed 19.11.15]

[2] 100Kin10. <https://100kin10.org/> [accessed 14.08.16].

[3] STEMconnector. <http://stemconnector.org/about-sc/> [accessed 14.08.16].

[4] Where are the STEM students? <http://www.discoveryeducation.com/feeds/www/media/images/stem-academy/Why_STEM_Students_STEM_Jobs_Full_Report.pdf/> [accessed 14.08.15].

[5] Hole-In-The-Wall Beginnings. <http://www.hole-in-the-wall.com/Beginnings.html/> [accessed 19.11.15].

[6] List of organizations engaged in STEM education across Africa. <https://en.wikipedia.org/wiki/List_of_organizations_engaged_in_STEM_education_across_Africa> [accessed 16.08.16].

[7] Master of Science in Medicine Degree Program. <http://msm.stanford.edu> [accessed 11.09.16].

CHAPTER 2

The Local Perspective

F.M. — So getting into graduate school is actually the easy part, isn't it? I remember seeing a graduate program staff member wearing a t-shirt saying "Now that you got in, I will help you get out." I laughed back then but I think it is a very fitting statement. The fact of the matter is that we will need help to not only get out, but to make it out in one piece.

J.T. — And ideally not just getting out in one piece but thriving, building resilience, and learning to become a truly independent scientist while still maintaining strong friendships, relationships, and wellness. A truly tall task!

F.M. — There is of course still more training to do but you ideally want to come out a CONFIDENT and independent PhD holding trainee scientist.

J.T. — Oh confidence. Confidence is tough in graduate school because so much of what I experienced was confidence shattering! Did you ever feel the same way?

F.M. — Certainly. I think it was the same for many people. We were all certainly shaken and our confidence was tested in ways we never experienced before. I think that's the whole point.

J.T. — Isn't it ironic that so few of us know this going into graduate school? And what do you think confidence means? I think it takes many forms.

There is confidence in self: that I know how to design experiments, actually perform the experiments, and analyze the data.

There is confidence in data: that I've repeated something enough times, compared to the appropriate controls, thought of all the possibilities.

There is confidence in the advisor: that he or she will support you throughout your career, that their decisions are in your best interest, and that they will financially support your work.

There is confidence in the process: that you will, at some point, make it out of the program as a fully minted PhD scientist. Phew!

Journeys in Science. DOI: http://dx.doi.org/10.1016/B978-0-12-813090-2.00002-6

F.M. – Well said! I think the confidence in self tops the list for me. I wanted to have the confidence to say I DON'T KNOW.

J.T. – What do you mean? That you felt like you always had to have an answer?

F.M. – Sort of. It has to do with how one feels when you have to admit that you don't know the answer or you are afraid of getting it wrong. I do feel that an attitude of curiosity is lost as we get conditioned in the traditional educational system. The process of discovery begins with having to admit that we do not know.

J.T. – I know what you mean. I always felt so guilty saying "I don't know." It just always seemed like the "wrong" answer. You are absolutely right that mature thinkers are able to say "I don't know" because they are really able to assess their own knowledge gaps. Now, I feel like knowing my own weaknesses is often more important than recognizing my own strengths.

F.M. – It took me a while but I finally got there. It was a process. I had to overcome what I call my negative committee. I mentioned this earlier but I think we can explore it further here.

J.T. – Ah yes, what exactly is the negative committee?

F.M. – A negative committee consists of voices that essentially challenge pretty much what your own true internal voice is trying to say. One member may say "I don't think you are ready for graduate school." Another may say "you don't belong here." Another may even suggest that you just keep quiet. The list is endless. My favorite is "you will be found out." Of course, all these can come from one committee member inside your head. They are essentially in charge, and the unfortunate thing is that some of us actually listen to them too much of the time.

J.T. – The negative committee is such a tough thing to face but you provide such a vivid description. I can still hear all the members of my negative committee telling me that I wouldn't be able to do a particular experiment, that I would never finish one degree, let alone two, that I would mess up a talk or a presentation. The list goes on. When did you first recognize this?

F.M. – Sadly, I fully noticed this in my last year! Yup. It took me that long. I somehow managed to fight them off throughout graduate school not realizing who I was really fighting. I think it has a lot to do with the self-reflection I did toward the end when I questioned whether the light at the end of the tunnel was a way out or a train barreling down the tunnel. I would encourage readers to explore the article I wrote as I think it captures what I felt very well at that time [1]. It is striking reading it now. I admit that I don't think everyone has a negative committee. Of course, some of us have it worse than others. The important thing is to realize that the committee can exist and that you don't have to listen to its members ALL THE TIME.

J.T. – At least you recognized it! And not only that, you made so many others aware of their own negative committees.

F.M. – I was humbled by the large number of responses I got from people.

J.T. – I think your article helped so many people realize they aren't alone. There was always an element of graduate school where I felt like I was completely alone. At the end of the day, I was responsible for my own successes and failures. I was the one driving the experiments forward. If I didn't work during the evening or on the weekend, it was on me. I was the only one I could blame. And on top of it all, many of my friends had started working real grown-up jobs instead of pursuing graduate school after college. And it was very hard for me to relate, and vice versa for them to understand what I was going through. Did you ever feel this way?

F.M. – Not quite. Most of my friends actually ended up going to graduate school right after college as well so most—I would like to think—understood what I was going through. Of course, I had to explain to my parents from time to time about some of the things I was going through in graduate school and the work I was doing. But they mostly understood though.

J.T. – That's great that many of your friends understood what you were going through. What strategies do you have for actively fighting off your negative committee? Honestly, what I found to be most helpful was failing well. Failing miserably at something. Sometimes it was super silly, like spending tons of time collecting DNA samples, running my PCR, loading the gel, and then DROPPING THE GEL ON THE GROUND as I walked to the gel box. When I fell flat on my face, I learned that (1) I was actually totally fine, (2) everything was going to be okay, and (3) I was less afraid of falling the next time.

F.M. – That's funny.

J.T. – I swear every biosciences PhD student must have dropped a gel on the ground at some point...

F.M. – I totally dropped one! But on a serious note, failing is important. There was actually a whole project that did not work out in my hands. The preliminary results were initially promising but it did not work out in the end. I had spent quite a bit of time on it too. Thinking back, my negative committee was at its loudest during that time. I think what helped a little during this time was the community of friends I had around me. I was involved in the graduate student council and was the chair of the diversity committee. That helped a little. Furthermore, I also had a small group of friends with whom I shared my woes with. In time, I came back to the bench more determined and focused to prove the negative committee in my head wrong. I think it speaks to the importance of resilience in graduate school. You will need lots of it. Heck, one needs it in life.

> *J.T.* – I could not agree with you more! Failing was and is important. From my own experience, I found that graduate school is all about bouncing back. And not just being okay but really coming back with a fierceness. Like someone lit a fire under your rear end. It's all about being voracious.

Sharing experiences is so essential. I found that not only in the sciences but in other disciplines—Education, Business, Law, Philosophy—my other graduate student friends were going through such similar experiences. Living in graduate student housing also gave me a strong community where in the evenings, I could grab a meal with someone and just chat about our days, debrief what was going on, and support one another. Inevitably, there were some days where I was having a great day and someone else was having a terrible day, and vice versa. But, graduate school is truly where I gained resilience. What do you think are the best ways to nurture resilience? Because I think not everyone gains resilience; some let the failure eat them alive. It can be very debilitating.

F.M. – That is not an easy question but I think the key lies in mentorship. I think we have said this multiple times already. It's a given that we will experience difficulty in life. Graduate school is no exception. Having the right mentors is crucial.

J.T. – So true. I remember discussing this with you multiple times during graduate school, but I think one of the most important lessons I learned was not to worry about things that are out of my control. What a tough lesson, but one that really made a huge difference for me and decreased stress related not just to graduate school but to life!

F.M. – Can you talk a little more about that?

J.T. – This was a surprisingly tough thing for me to learn during graduate school. The simplest example I can think of is submitting a manuscript, fellowship application, or grant application. Once I submit everything, there's really nothing left for me to do. The decision on whether my paper goes for review or I get a particular fellowship is no longer up to me. The natural response is to worry about it. Worry about whether I should have changed that one figure legend or whether I should have switched the order of Specific Aims 1 and 2. But, I found that I was spending a ridiculous amount of time worrying myself silly about these things. And it wasn't worth it. I was getting worried sick. So letting go and being okay with uncertainty is something that has not come easily. And I am still learning!

F.M. – I remember having the same feelings when I submitted my first paper. It was quite silly as I think back now. I should have enjoyed more the idea that I was submitting a paper in the first place. Of course, it's easy to think this years down the road. Anyway, back then, I was worried about a lot of little things. But you know, the one thing that is also out of your control is how people (one's advisors, peers, etc.) think about you. It is something I think many of us worry about a lot. Obviously, this is not just a graduate school thing. It's life but many of us forget this during graduate school. The criticisms can slowly degrade one's confidence if you allow them. But let's face it. Someone will eventually tell you that your presentation sucked. Someone will eventually tell you that you need to take a writing class. Someone may even tell you that they think you are quite simply wrong!

J.T. – There's something very powerful about being able to let go. Oh yes, worrying about what others think of you is something I grappled with a lot. One of the best pieces of advice one of my mentors gave me (repeatedly) was not to take rejection personally. We put so much effort and time into our experiments, manuscripts, and grant applications. I find myself getting very emotionally invested that when these things get rejected (and the majority of time, more often than not, they will get rejected!), I take it as a personal hit.

F.M. – Exactly. Don't take it PERSONALLY. Gosh, we take a lot of things personally.

J.T. – Precisely! We sure do take things personally! How did you stop all of the criticisms from getting to you? Even though I really actively try to let criticisms bounce off of me, there are times when I replay criticisms over and over again in my head.

F.M. – I will be honest. I internalized many of the early criticisms in graduate school. I freaked out and got defensive when someone gave me some criticism. I always feared of getting an external confirmation of what my internal negative committee was saying to me. It was and, I guess it still is to some extent, an interesting dynamic.

J.T. – Gosh, I went through the exact same thing. Someone would just make one small comment, and I would perseverate on it for days. I think I've definitely developed a thicker skin now. And, honestly for me, one of the most important things at the end of the day is knowing that I've done the best that I can. I can be satisfied with that, no matter the outcome that someone else dishes out to me.

F.M. – It is totally about building that thicker skin. Of course we all react differently to criticism. I think one needs to accept that it is part of the learning process and not to take it personally. The last part is difficult.

J.T. – Resilience is the key. In the past, when I got a rejection, I would think about everything that I did wrong. I would look into all my prior actions and think to myself, "Why did I do X? Why didn't I do Y?" It's all about your frame of my mind. Now when I get a rejection, I just let go of the past. There's nothing I can change about the past. What I can change is the future. And if I let go of that, it makes it easier to be not only a productive scientist, but a mindful person!

F.M. – I actually like rejections now. One grows a lot more from rejections. In the end, I think trainees need to get used to negative criticism. It is the nature of academia and, of course, life. Given the funding climate now, for example, there are a lot of rejections out there. A thick skin is absolutely necessary.

J.T. – Very true. The flip side (that we rarely think about) is also important. The postdoctoral fellow who I sat back-to-back to in graduate school used to always remind me that it was important to fully celebrate and appreciate people's successes in the lab. If someone had an awesome experiment finally work or if a manuscript got accepted, it was so important to acknowledge and enjoy that! I always really liked that philosophy.

F.M. – Celebrate the small things too! The little wins do matter. Many of us forget this.

J.T. – The small victories are important! And I think it's important to share that with your scientific community and amongst colleagues. What do you think the best approach is for mentors in handling failure and rejection in graduate school? What makes this tough is that, in some ways, graduate school is a series of repeated failures, one after another. Continuous failure can be really overwhelming, and how your advisor and mentors respond to that is key.

F.M. − I think it is helpful to tell mentees to expect failure and that they should use such experiences to grow.

I do think too much hand holding is bad. Graduate students need to go through the many pressure tests of graduate school.

J.T. − I think it's all about balance. Too much hand holding doesn't allow the student to gain independence but ignoring the difficulty of it all is also not productive. I think it's important for the mentor to acknowledge that failure stinks but to also push the student to bounce back from rejections and to push forward. Do you remember the inflatable bounce houses?

F.M. − Of course!

J.T. − I always imagine failure like that. You just have to bounce back!

F.M. − Of course, trainees need to also use resources outside the lab.

J.T. − Definitely! What kinds of things did you do outside of the lab?

F.M. − I think it is crucial to avoid isolating oneself in handling rejections or after going through difficulties in the lab. I found the time to get involved in the graduate student union and sought other opportunities to participate in university life. I think it totally helped. I think sharing your experience with others who are likely going through the same thing helps put things into the right perspective. It is a gentle reminder that you are not alone.

J.T. − Absolutely. Isolation is dangerous.

F.M. − Very.

J.T. − I really enjoyed being active. I got really into exploring different hiking trails in the Bay Area. There was something exhilarating about being outside, and I would make it a point to turn off my cell phone—no emails, no phone calls, no texts—something about that would really put things in perspective for me.

F.M. − I actually found myself running a lot.

J.T. − You've run many more races than I have! There is something funny about running that seems to parallel graduate school. Every time I do a race, I tell myself I'll never do one ever again. There's this sense of feeling simultaneously exhausted and elated. And then, somehow, I find myself signing up for another one.

F.M. − Hahaha. What parallels do running, say a half marathon or full marathon, have with graduate school do you think?

J.T. – So many! First off, there's a lot of training and preparation that must be invested up front even before the race itself. In graduate school, it's the same. Tons of reading, literature searches, you have to pass a qualifying exam. It's like qualifying for the Boston marathon. You have to pass one hurdle before you can move forward. What do you think?

F.M. – Hmm, I totally agree with that but would you suggest—say to an undergraduate reading this—run at least a half marathon before applying to graduate school? Perhaps they should make it part of the application. That's one solution I have not heard anyone mention in combating the oversupply of PhDs that everyone is talking about!

J.T. – Hahaha. Must run half marathon prior to application cycle.

F.M. – Exactly. There would be a revolt.

J.T. – Must complete Ironman in order to graduate.

F.M. – Okay, okay. I think we have taken it too far now. On a serious note, resilience is important, but what happens when one cannot handle it? We have all heard of such instances.

J.T. – Yes yes, in all seriousness. This is really tough. Mental health is a huge issue in graduate school, particularly for students in STEM fields.

F.M. – But we actually don't know the full scope of the issue. There is very little data out there right?

J.T. – It's true. There are bits and pieces of data on graduate students, and virtually no data on postdoctoral fellows [2]. The data that does exist on graduate students is, quite honestly, frightening, to say the least.

F.M. – The University of California at Berkeley (UC Berkeley) is leading the pack in detailing mental health in graduate school. Its most recent report states that the top predictors for graduate student well-being include financial stability, career prospects, advisor relationship, overall health, and social support [3]. The usual suspects.

J.T. – I would argue that the UC Berkeley report is one of the best we have. They did a full, comprehensive survey of the entire graduate student population in 2004 [4]. And they did this not only once but actually did a follow-up study about 10 years later in order to assess changes. How did these top predictors affect you during graduate school?

F.M. – I think the advisor relationship was crucial for me. I am thankful that I had a great relationship with my advisor. It made a difference.

J.T. – I'm thankful because I felt well-supported on the graduate student stipend I got. I can't imagine how difficult it would be to have a family and have to think about how to pay for all the costs associated with having children.

F.M. – This is very true. The financial situation varies considerably among trainees at different institutions and cities.

J.T. – It just adds a huge layer of stress and complexity to an already stressful experience as a graduate student.

F.M. – True. I remember reading in the UC Berkeley report that the number of students reporting mental health service utilization was actually lower than those reporting need. Do you remember this?

J.T. – Yes—it forces you to think—WHY? An easy assumption is that people don't get services because services don't exist. But the situation is more nuanced than that. There is definitely stigma associated with utilizing traditional mental health services like counseling. What other reasons do you think there are for poor utilization? I don't think this is very well studied.

F.M. – I think lack of awareness is another issue. Some just don't know. There are probably many nuanced reasons. You are right that this is an issue that needs more study. We can't just assume everybody is alright.

J.T. – Going back to what you said earlier, I agree that a healthy advisor relationship is key. I am so grateful that my advisor cared a lot about my work and the science that I did, but also really took a vested interest in my personal life and my well-being. That certainly is not true of all advisors, unfortunately.

F.M. – It is very important to know how to manage advisor–advisee relationships. It is a delicate balance.

J.T. – What advice would you give to undergraduates and new graduate students in terms of managing the advisor–advisee relationship?

F.M. – Take the initiative. I do think it is critical that mentees take the initiative to meet their mentors more than half way down the road. Don't just wait around for your advisor to summon you. Take the initiative to ask for meetings and try to even schedule them on a regular basis. Show your advisor you are taking the relationship very seriously. I have certainly dropped the ball on this a couple of times. It is very easy to just wait on the advisor.

J.T. – I couldn't agree more. I think it can be very intimidating initially. In some ways, there is an inherent power dynamic built into the advisor–advisee relationship. The advisor is a faculty member, has more experience, and often dictates whether or not you can graduate!

But, you're right, the advisor should be expected to look out for the student, but the reverse is also true. It goes both ways.

F.M. — This is part of the reason why I see advisors very differently from mentors.

J.T. — Ah, interesting distinction. How so?

F.M. — I think mentors take more of an initiative to reach out to mentees. They tend to be equally, if not more, invested in the relationship. They may even advocate on your behalf. You can have lots of advisors but few mentors.

J.T. — I never thought of there being a nuanced difference between the two. So, a mentor is certainly an advisor. But, an advisor is not necessarily a mentor?

F.M. — That's right.

J.T. — One question that I often hear from many undergraduates and folks considering graduate school is: How many mentors should I have?

F.M. — It's about quality and not quantity. I would rather have a few good mentors rather than a bucket of advisors. In fact, it can get a little complicated to manage a large number of advisors. Figuring out whose advice to follow could be tricky.

J.T. — Yes! One of the most difficult things is learning how to filter advice. Everyone will tell you their opinion if you ask them, but you have to sift through and weigh each piece of advice accordingly. I've found that I have different mentors for different realms or spheres of my career.

F.M. — It is certainly helpful to have mentors for different realms. It's interesting that mentorship keeps coming up everywhere. I have a hint it will keep coming up.

J.T. — Definitely. Mentorship is the name of the game. I definitely have physician—scientist mentors as well as mentors specifically within Pediatrics, others specifically interested in Neuroscience, and others yet who advise me in writing and various advocacy projects. So, here's a tough question. What do you do when you disagree with an advisor or mentor?

F.M. — This is where it helps to have another advisor or mentor on hand. As a graduate student, one typically has a committee. A student can reach out to one or two of the members and get their take on the issue.

J.T. — Definitely, I also remember in graduate school, it was always nice to bounce ideas off of other students in the lab or other graduate students in your program. I've found Program Directors to be great advocates for students as well.

F.M. — They totally can be!

J.T. — Also most institutions have an ombudsperson who is a neutral person who can give advice in confidence.

F.M. — Such a resource is important to know about and take advantage of when necessary. In my experience, there are always resources lurking around but the issue is not knowing about them. For example, an issue that becomes important over time is work–life balance. I remember Stanford's Office of the Vice Provost for Graduate Education had great resources for this.

J.T. — Definitely, sometimes there are so many resources it can be hard to know where to start. I also found Stanford's Graduate Life Office to be a great place to get good advice and support!

F.M. — Of course we are biased towards Stanford but many institutions have similar resources.

J.T. — Yes, we were lucky to have so many resources at our disposal. What do you think about the concept of managing up? What exactly does that mean?

F.M. — Managing your advisor or mentor.

J.T. — Seems counter-intuitive, no?

F.M. — True. The idea is to help your advisor or mentor understand what you need.

J.T. — Ah yes. I like to think of it as being empowered to go for what you want!

F.M. — I can hear people asking, "How do I do that?"

J.T. — The foundation of being able to manage up is having solid communication with your mentor.

F.M. — You have hit the motherload. It is essentially about communication.

J.T. — If this is the first time that you're asking anything of them and you never communicate... then that clearly will not go over well!

F.M. — Something as simple as having a typed out agenda for meetings could be used to help your advisor or mentor understand what he or she needs to do for you. You can even send it in advance if possible.

J.T. — Totally! That is huge. Seems simple but you'd be surprised. Just being organized I think is appreciated by mentors tremendously! Although this is certainly easier said than done. Sometimes, others have their own agenda for a particular meeting, and you can easily become sidetracked or derailed. Another important part of managing up is a discussion of long term goals.

F.M. – You mean those SMART goals!

J.T. – Yup!

F.M. – Working with your advisor or mentor to put together goals that are specific, measurable, attainable, relevant, and time specific early in one's training can be very useful.

J.T. – Well said. What were some goals that you set?

F.M. – The department had very specific milestones and so we kind of followed the map they drew for my specific track. It was certainly helpful.

J.T. – Nice. I always had both short-term and long-term goals. Short-term included immediate experiments, papers I wanted to read, specific fellowships I wanted to apply to. Long-term included much more far-reaching goals like what I wanted to specialize in medically and what I wanted my long-term career to look like.

F.M. – I think such discussions vary so much from advisor to advisor. We certainly talked about these things at some point but I just don't remember the full details which is interesting now that I think about it. Perhaps it was working so well that I didn't have to think about it.

J.T. – That's awesome that it came so naturally!

F.M. – But I do think an individual development plan (IDP) has to be in place somewhere because it is easy to forget to do such things.

J.T. – Agreed. And it needs to be constantly readdressed, revisited, reassessed. It's an iterative circle.

F.M. – Exactly. I would advise trainees to continually ask themselves, "How can this be better? What can I do better?"

J.T. – Ah yes. Always improving, changing, and taking new data into account! As any good scientist should. Did you ever have this feeling that you were "bugging" your advisor?

F.M. – I would guess a high percentage of mentees feel like that. Of course, if your advisor's door is closed, then you totally feel that you are bugging your advisor.

J.T. – Hahaha yes. I think this is sort of difficult to explain but there were certainly times early on in graduate school where I was really nervous to ask questions. But as I progressed through training, I got much more comfortable "bothering" my advisor about random things, sometimes related to science and sometimes not.

F.M. – It really depends on the relationship one has with the advisor. Of course, all these things add to a mentee's committees in his or her head.

J.T. – I think it is important to remember that one negative interaction or failed experiment does not mean that the relationship with the advisor is over. In the moment, it's hard to remember that. I remember the committees in my head saying, "I wonder if this professor thinks I'm a good scientist?" "I wonder if that professor will even respond to my email?"

F.M. – True. True. There are both internal and external forces influencing graduate students. We can name a lot of them. Going back to mental health, there needs to be a greater awareness that this is an important issue to address.

Now, I can hear some critics—or maybe my own negative committee—saying that there is a risk that we are trying to make graduate school easier. "Graduate school is supposed to be difficult. It is supposed to push you. I went through it just fine. If you can't handle it, get out." What would you say to them?

J.T. – This is a tough one. I would say that there is some truth to this. But, there is also evidence that some graduate students are really struggling, very seriously—including paralyzed by anxiety and contemplating suicide [5]. These are serious issues and brushing them off as "the way things are" is missing the point entirely. Everyone goes through ups and downs—it's a natural part of life—and during the tough times, we need our communities to really support us. What would you say?

F.M. – I completely agree with that. The issues facing some graduate students are real and we should all be concerned. International students are one group that I think deserves mention since a large chunk of graduate students come from abroad. Being so far away from home, worrying about finances, research, and the rest of it can be overwhelming. Minority students are also another group. I think support structures do exist in many places but some still fall through the cracks.

J.T. – International students are absolutely an important group. There has been some data showing high rates of anxiety and depression amongst international graduate students [6]. I think there are also significant cultural differences when some students come to the United States that are quite different from their home country. Language barrier can be a tough one as well.

F.M. – Of course not all international graduate students face such issues. At the same time, we shouldn't assume, as I mentioned earlier, that everyone is okay. Anyway, I think another group that we don't hear much about are those that pursue the combined MD-PhD degree. I can't imagine what it's like for them.

J.T. – Ah yes, we sort of exist in this limbo world where we simultaneously straddle the world of medicine and the world of science. It's a delicate balance, for sure.

F.M. – You do a nice job talking about it in your Science Careers article [7]!

J.T. – Thanks!

F.M. – What inspired it?

J.T. – Honestly, throughout graduate school I felt like a double agent.

F.M. – I see.

J.T. – I felt like my clinical and scientific obligations were always fighting each other for my time. I was constantly being pulled in one direction, then pulled in the other, back-and-forth like a tug-of-war.

F.M. – Is that stressful?

J.T. – Definitely. There was this constant feeling of not having enough time. That I should be studying for a medical school exam but then I should also be doing an experiment.

F.M. – I just can't imagine having to do both degrees. How did you manage?

J.T. – You know, for me, it was really important to maintain perspective. To remember my origin story. My patients actually really inspired me the most. Taking care of sick patients really kicked me into gear. It reminded me of why I was doing what I was doing—so that I could not only take care of these amazing people but so that I could also help them through my research.

F.M. – That makes sense. What advice do you give undergraduates interested in the MD-PhD? I have met a number of them interested in the track over the years.

J.T. – MENTORSHIP is the answer to everything!

F.M. – It totally is.

J.T. – Hahaha, in all seriousness, finding mentors is key. Do your homework. Talk to actual MD-PhD students at the end of their training. Meet real physician-scientists and learn what their day-to-day work is like. I was lucky to have a physician-scientist mentor during undergrad who guided me through the process.

F.M. – And how should they think about the stress of it all? What sort of attitude should they have?

J.T. – Great question. I think the duality of the physician-scientist career is actually one that I find to be super rewarding. To scientists, I am an ambassador of medicine. To physicians, I am an ambassador of basic science. My attitude is that there will always be more to learn! And rather than being overwhelmed by that, I'm thrilled by it! It's an adventure of exploration, learning, and ultimately the point of it all is to bring science and medicine together to improve the lives of patients. It's a truly unique position to be in.

F.M. – Indeed. However, like other graduate populations, I do think the MD-PhDs also deserve a close look at their mental health.

J.T. – Absolutely. Institutions need to do their due diligence and assess the mental health needs of all their student populations. By the way, have you ever heard of the Blue Zones?

F.M. – I have not.

J.T. – Blue Zones are fascinating to me [8]. There are five identified blue zones—Icaria, Greece; Loma Linda, California; Okinawa, Japan; Nicoya Peninsula, Costa Rica; and Sardinia, Italy—they are areas of the world with incredible longevity.

F.M. – What do they owe their longevity to?

J.T. – Interestingly, they are all populations in which community and family are the core focus of life. Physical labor is a part of daily living as well—they get high amounts of physical activity. Every member of the community plays a critical role in the social structure.

F.M. – What are you suggesting?

J.T. – There is something incredibly important about community— feeling like you are contributing to a larger good and being a key member of a team.

F.M. – Like you are contributing to something bigger than yourself.

J.T. – Exactly.

F.M. – This is a powerful idea I think few understand.

J.T. – People in these communities have a very high level of civic engagement—caring about the people in their communities, the infrastructure, the policies. I think sometimes in the laboratory, we can get siloed from other people and so narrow-minded that we forget about what's going on around us. I've definitely been guilty of that.

F.M. – I think many of us have been guilty of that at some point.

J.T. – I think an important part of graduate school was just hanging out with other graduate students, supporting each other, going to happy hour, grabbing a meal between experiments, or having a coffee break during the day. Did you think that helped you during graduate school?

F.M. – It ALL helped.

J.T. – Definitely. Supporting peers is key but are there higher level approaches that could be taken?

F.M. – I think something can also be done at the level of the NIH. Maybe create an office or something.

J.T. – Very interesting, what do you envision that office would do?

F.M. – I imagine the office would provide some top-down guidance in setting a strategic vision to expand support for mental health and wellness nationwide across the entire biosciences community. As the largest funder of research in the biosciences, the NIH is perfectly situated to improve the mental health climate.

J.T. – I really like that idea. I think sending a strong message from the top about the importance of mental health would make huge waves.

F.M. – I remember you mentioning an innovation fund or something when we discussed this a while ago.

J.T. – Yes! I was envisioning a mental health innovation fund that would encourage creative mental health interventions and help to fund out-of-the-box projects meant to help graduate students.

F.M. – A cool idea. I do think things will change for the better in the future in graduate education. Future graduate students will have access to so many resources to do the best science possible. At the end of the day it is really about the science.

J.T. – You're right on point. We HAVE to address mental health because it affects our ability to do great science and reach our highest potential.

F.M. – I think this is critical. We always have to be thinking about the science and asking ourselves the following question: how can we do BETTER?

J.T. – Always striving to be better, always improving.

F.M. – It is this attitude that I employ when I also think about science all over the world especially in developing countries. Anyway, we will talk more about this soon.

Something that also troubles graduate trainees is the continued discussion about the lack of positions in academia. I mean it's everywhere. Today's graduate students (especially those in the biomedical sciences) are becoming well aware that there are simply too many of them being produced with respect to the number of open positions in academia. I am guessing that many are probably asking themselves, "What else can I do? Do I have necessary skills to move outside academia? Which industry fits me best? Will I have the support from my department and research advisor in pursuing jobs outside the academy?" These are probably just a few of the myriad questions they face.

Given that many won't end up in academia, one interesting question then is how should graduate programs position themselves to train the next generation of students?

J.T. – Ultimately, I think we have to prepare trainees for the reality of the academic environment, which means exposing them to industry, journalism, policy, advocacy, start-ups, data science, and more DURING graduate school. The issue is obviously that most advisors are typically academics, and therefore as a student, you have to make a concerted effort to gain exposure to what is out there beyond academia.

Do you think most prospective graduate students are aware of the lack of positions when they are considering graduate school and just starting off? Or is it something they find out once they enter graduate school?

F.M. – It depends on the department but I would venture to guess many do not know. They are mostly concerned with just getting in! It is part of the reason we published an article to help trainees know where the nonacademic resources are [9].

J.T. – What I worry about is often, I think, some new graduate students begin with the expectation that academia is the only path. I wonder how we can expose students early on to other pathways. I also wish there was a better word than "alternative" career path. It makes it sound like academia is the primary path and that anything else is because you "couldn't cut it."

F.M. – I strongly believe that scientific training needs to remain a central focus in the PhD training. We need more scientific thinkers out there! Now, I also think we need more friends of science in society. We need people trained in the sciences to help make good policies that support science in Washington and around the world. We also need science writers who are able to communicate research to the masses. I can keep going.

J.T. – Oh yes, I agree completely. But, I think we are missing some bits and pieces. We really desperately need outstanding scientists who bridge multiple worlds. In some ways, this reminds me of physician-scientists. But we need journalist-scientists, politician-scientists, policymaker-scientists, lawyer-scientists, CEO-scientists, photographer-scientists, filmmaker-scientists, advocate-scientists, and more. Is graduate school the best time to be exposed to this? Or later? Or earlier?

F.M. – I think graduate school is the time. Of course, some have argued that it would be better to maybe create a professional PhD for those whose interests lie outside academia and also have an academic PhD for those interested in academia. It is similar to what is done in engineering I think.

J.T. – Hmm, I had never heard of that before. Interesting concept though. I still think that formal academic training is important for people who go beyond academia. That kind of rigor makes them able to bridge the academic and nonacademic worlds.

F.M. – It is an interesting debate. The idea is such that trainees in the academic-track PhD would focus on blue-skies research and fundamental discovery whereas a vocational PhD would be more structured and directed towards other careers.

J.T. – Are any institutions trying this yet?

F.M. – I am not sure actually. I can imagine the academic PhDs looking down on the "other" ones. This split PhD idea is something Julie Gould discusses in great detail in her Nature article [10]. It is definitely worth a read.

J.T. – Yeah, I worry about that. Did you explore nonacademic career routes during graduate school? What kind of exposure did you have?

F.M. – Interestingly, I actually wanted an academic position. I explored the nonacademic tracks only in my head. I am not sure what I thought that would achieve.

J.T. – I actually dabbled a bit. I did a one year Graduate Public Service Fellowship at the Haas Center for Public Service during my PhD [11]. I was really lucky that my advisor supported this.

F.M. – Support from advisors is instrumental. How did you negotiate this with your advisor?

J.T. – Honestly, I just went to his office one day with the description of the fellowship and the application. And I told him that it was something I really wanted to explore. And I think he needed to write me a letter of recommendation. It was all very boring and straightforward. I think what made it easy was that my mentor and I had discussed that public service and advocacy were things that I really valued as part of my career. I was also making good progress in my PhD. I think if I had been behind on experiments, he may have thought twice but I can't say for sure.

F.M. – I think it's great your advisor supported you on that.

J.T. — The fellowship was a really interesting concept. I hope other institutions adopt it as well. The fellows were from all different departments across the university including Psychology, Education, Engineering, Law, English, Anthropology, Economics, Sociology, and more!

F.M. — What was the core idea behind it?

J.T. — The core idea was to engage graduate students in public scholarship and service, in the context of academia and outside of it. We had incredible weekly seminars taught by both academics and non-academics participating in civic work that truly engaged the community at large. There was an emphasis on how to incorporate civic mindedness into an academic life. It was all incredibly eye-opening for me. And so wonderful to meet other graduate students from such varied disciplines who all shared a passion for public service, diversity, and leadership.

F.M. — I think I missed that memo. There are so many opportunities in graduate school! I sometimes feel I didn't take full advantage.

J.T. — Yeah, there are always so many opportunities! You certainly can't take them all, that would leave no time for lab work!

F.M. — True. Research is king.

J.T. — I knew several people in graduate school who participated in cool data science boot camps. Have you heard of that?

F.M. — Ow yes! It certainly takes a well organized trainee to be able to manage his or her research and still be able to do these additional, and of course important, activities. Summers are a really good time for them.

J.T. — That's a great time to explore! I would also encourage folks to go to seminars (not just science or engineering-related) around campus to stimulate more thought and discussion. At our MD-PhD program retreat one year, Sheri Fink who was an MD-PhD alum came to speak. She is actually an investigative journalist who has won the Pulitzer Prize [12]! Her work is so influential, and she has both substantive medical and scientific training. That was the first time I realized a scientist could be a journalist!

F.M. — I think we have explored so many topics! Of course, we are only still covering the local perspective. This is a small piece of a very big puzzle. It is very easy for us to forget that the scientific enterprise includes the whole world. There are trainees across the world who face many of the issues we have brought up. But there is a whole unique set of issues that trainees abroad face which impacts their ability to do their best science. It is critical those of us in the United States keep this in mind.

J.T. − Absolutely. There are a host of other issues to take into account once we take the global perspective. Hearing the varied and diverse scientific origin stories will only improve the way we approach STEM education. In the United States, we have a lot to learn by international collaboration and idea sharing.

F.M. − So, how about we zoom out a little. Let's change our perspective.

J.T. − Let's go!

REFERENCES

[1] Muindi F. Tell the negative committee to shut up. Science 2014;345(6194):350.

[2] Tsai JW, Muindi F. Towards sustaining a culture of mental health and wellness for trainees in the biosciences. Nat Biotechnol 2016;34:353−5.

[3] Graduate student happiness and well-being report, <http://ga.berkeley.edu/wellbeingreport/>; [accessed 21.01.17].

[4] Hyun JK, Quinn BC, Madon T, Lustig S. Graduate student mental health: needs assessment and utilization of counseling services. J College Stud Dev 2006;47:247−66.

[5] Garcia-Williams AG, Moffitt L, Kaslow NJ. Mental health and suicidal behavior among graduate students. Acad Psychiatry 2014;38:554−60.

[6] Hyun J, Quinn B, Madon T, Lustig S. Mental health need, awareness, and use of counseling services among international graduate students. J Am College Health 2007;56:109−18.

[7] Tsai JW. The M.D.-Ph.D. double agent. Science 2015;350(6266):1434.

[8] Blue Zones, <https://www.bluezones.com/> [accessed 21.01.17].

[9] Muindi F, Keller JB. Emerging network of resources for exploring paths beyond academia. Nat Biotechnol 2015;33:775−8.

[10] Gould J. How to build a better PhD. Nature 2015;528:22−5.

[11] Graduate Public Service Fellowship, <https://haas.stanford.edu/students/cardinal-careers/fellowships/graduate-public-service-fellowship/> [accessed 21.01.17].

[12] Fink, Sheri. The deadly choices at memorial. The New York Times, <http://www.nytimes.com/2009/08/30/magazine/30doctors.html?_r=2&/ > ; 2009 [accessed 11.09.16].

CHAPTER 3

The Global Perspective

F. M. – Understandably, the discussion about training in STEM is mostly focused internally. What I mean is that the discussion is centered on the United States. It is very easy to forget about the rest of the world. I think the developing world deserves special attention. It is an area I am always thinking about. Organizations such as the World Bank and UNESCO are leading the way in strengthening STEM infrastructures in developing countries. Of course, they are not alone. There are many other organizations doing some innovative programming in an effort to inspire and build capacity around the world.

J. T. – It's true, we often don't think about this at all and are very focused on what is happening domestically. I think the opportunity is ripe to not only inspire new scientists around the globe but to also put our minds together as an international community to determine what the most creative and revolutionary strategies are for improving STEM training. Where do we even begin? This is such a huge topic!

F. M. – It is a huge topic indeed! Let's start at the beginning. When thinking about developing countries, the typical problem is usually funding.

J. T. – I have so many questions to start off! Is there an NIH equivalent in other countries? What are the major funding sources? And for folks who don't have access to an academic institution, for example, how would they access funding to do their own research?

F. M. – All good questions! If we take Africa as an example, NIH equivalents do exist in some countries [1]. These include the Africa Research Excellence Fund, African Union Research Grant, and several others. Of course, their budgets are no match to the NIH in the United States. Even so, we must take a step back before we think about research funding. You see, despite recent progress, many parts of the developing world still face shortages of highly trained scientists and engineers. The 2010 UNESCO science report actually shows that African countries had an average of 164 researchers per million people in 2007, more than six times lower than the world average of 1081

Journeys in Science. DOI: http://dx.doi.org/10.1016/B978-0-12-813090-2.00003-8

researchers per million [2]. But to have a large enough pool of scientists, one must have a whole infrastructure in place.

J.T. — Wow, that is a frightening statistic. How do we create the appropriate infrastructure and support to boost the number of scientists in the developing world? Can you imagine the incredible innovation and literally world of ideas that could come from these new scientists? I guess the question really is—how do you build such an infrastructure and what are the key components? I think education is the essential building block.

F.M. — Exactly.

J.T. — It's the foundation to the house on which everything is built. What does STEM education look like in the developing world? I think a lot about how we can get children excited about science and engineering at a young age.

F.M. — It really needs to start early. Not only that, it also needs to be sustained over a long period of time. Lots of STEM organizations both here in the United States and in developing countries focus a lot of their attention on young people. I think this is the right approach.

J.T. — Early and often.

F.M. — I think the following word summarizes what STEM education looks like in the developing world: patchy. It is not sustained and, of course, there is the lack of access.

J.T. — So, ideally STEM education would start at an early age and be consistent throughout childhood, building on previously learned concepts. I think there is also a huge part of STEM that is currently missing which is laboratory and experiment-based science. This is a strategy, too, that can really draw kids in when they get their hands dirty, literally! Again, this is a resource issue given that obtaining materials, reagents, and glassware requires money.

F.M. — This is why I was so impressed with a nonprofit organization called TReND which stands for Teaching and Research in Natural Sciences for Development in Africa [3]. The group is run by volunteer researchers at several universities worldwide. It organizes outreach courses and workshops for young African scientists on how to conduct quality, affordable neuroscience research in resource-limited settings. In the 2010−2014 period, more than 1000 African students took part in their programs. They continue to do amazing things.

J.T. — Wow, that is incredible! Is TReND focused on college students? The real question then is how do people find out about the program? Getting the word out is key!

F.M. – They do the whole spectrum all the way up to college professors. They have a powerful local volunteer network which essentially gets the word out there.

J.T. – Wow, that IS incredible! This could potentially be a reproducible concept that could be applied to other countries as well.

F.M. – I think they are slowly spreading across several African countries. They really have the potential to make a difference. Of course, it is great that they are a little biased toward neuroscience.

J.T. – We are not biased toward neuroscience at all! :) So, we know that people participate in their programs. The next logical step as a scientist is to ask, do the programs make a difference? What are the outcomes of these students? More data!

F.M. – It is all about the data! The ultimate question is exactly that. What is the impact? What happens to the students? There are a growing number of STEM initiatives in Africa (remember there are like 6000 or so in the United States alone!). There needs to be some accountability put in place. We need to know who is doing what and whether it is having an impact.

J.T. – Could you imagine a collaboration across multiple countries that essentially involved a huge repository of this data? I imagine it could be public access so that anyone in the world could analyze it and assess the information.

F.M. – This is something my wife and I advocated for in an article back in 2014 [4].

J.T. – Yes, I remember—an awesome article!

F.M. – It was well received. In our research, the level of coordination between STEM programs wasn't clear. I am talking about Africa here. I think the situation is still the same. A central database to strengthen the landscape of STEM programs would go a long way in facilitating the implementation and spread of knowledge on what is and is not working. Of course, this can be applied at the global level as well. Maybe a Global Fund for STEM Programs or something similar.

J.T. – Yes, this would be awesome. The public access component is something I actually really value. I learned a lot about community-engaged scholarship through a fellowship I did during graduate school—this concept that if you are studying a particular community, you should engage them in analyzing the data and really involve them in the work. You get community buy-in plus it's a refreshing way to teach people about something that is directly applicable to themselves. The tough question is—for a Global Fund like this, who would manage it? Tough . . .

F.M. – That's a little tough. In the article, we suggested that this would be something that the various stakeholders would have to decide. I think a lot of insights can be garnered from the inception of the Global Fund to Fight AIDS, Tuberculosis and Malaria [5]. The process is critical.

J.T. – So, we know there are many groups working on initiatives. But what about within formal schools like elementary, middle, and high schools? This is where infrastructure comes into play to really provide access to STEM education to a large number of children in the developing world.

F.M. – The teachers within these settings hold all the keys. I think having quality teachers at each of these stages is likely to make a difference. We need to find a way to recruit the best science teachers to the classrooms.

J.T. – And ultimately, the core of this issue is not only recruitment but retainment. And I think the key to retainment is really a value issue where society needs to value the role that teachers play in child development. The bottom line is not only getting teachers, but providing them with the resources, support, and continued training to keep doing what they do best at a very high level.

F.M. – Here is an interesting idea. Curious to know what you think:

The recruitment and retention of high-quality teachers is one of the central tenets of many educational systems around the world. Some of the strategies used by many countries to maintain their stock of high-quality teachers include better monetary compensation, continuing professional development, leadership training, and also performance-based increases in pay. However, many countries—especially those in the developing world—continue to face shortages of high-quality secondary education teachers in several critical areas including science and mathematics. An additional approach that developing countries could use to attract the most talented science students into the teaching profession is to employ more media campaigns across television, radio, and the Internet. Such campaigns must show how important science teachers are for the future, and the important place they occupy next to other respected professions in society. Overtime, such active exposure will help (1) inspire more young students to start thinking earlier about a career in science education and (2) highlight the importance of science teachers in society, which will in turn help slowly elevate the status of science teachers across all education levels. The success of media campaigns will require coordination among schools, academic institutions, nongovernmental agencies, foundations, and also the federal government to ensure that the message is consistent. Of course, no single strategy is sufficient by itself as a number of factors govern the sufficient recruitment and retention of high-quality science teachers. In the end, I think media campaigns should be used as one part of a wider strategy in tackling science teacher shortage.

J.T. — Super interesting concept! It's funny, media exposure is so aggressive for things like movies, fashion, and accessories. But, we sadly don't use it for education. A logistical question is how many people actually have access to these media outlets. And what would these campaigns look like? I can imagine really impactful stories from teachers and students.

F.M. — I imagine an alternate reality where science teachers (actually all K-12 teachers) are paid salaries that are as high as engineers and perhaps even doctors. Why not? Wait, I think this is already a reality in some countries.

J.T. — Absolutely. Increasing salaries is not only a way to retain teachers but to demonstrate to society at large that their profession is one that we value greatly.

F.M. — Anyway, what would these campaigns look like? Nothing elaborate really. One simple strategy could be to just highlight the good work teachers are doing in the classroom. News media could do special highlights of teachers in specific schools. Why cover only the bad stuff happening in schools? There is a lot of good happening and this needs to be shown as well. It really is that simple. I think this is especially pertinent in developing countries where they have a unique opportunity to embed a culture that supports the teaching profession.

> Let the young kids see society take a special stance for teachers just like it does for other professions.

J.T. — Agreed absolutely. The culture needs to change to really attract incredible people to become STEM educators. And yes, there is a strange bias in the media to portray the ugly, the violent, and the unthinkable. It makes us falsely believe that only bad things happen in our world.

F.M. — There is so much untapped potential out there in young people. We just need to realize that we are much stronger together!

J.T. — Sharing is caring. That's what my preschool teachers used to say!

F.M. — I think you are right that the bottom line is to provide both teachers and students with the resources, support, and continued training to pursue science at a very high level.

J.T. — So, how do we build a country's capacity to pursue high-quality STEM? We have to start with funding. Where will the money come from?

F.M. — I think governments need to chip in quite a lot. Of course, the private sector has a role to play in this too. The most recent UNESCO Report cites the need for governments to raise their gross domestic expenditure on research and development (GERD) to the target level of 1% of GDP [6]. Interestingly, African countries with high GERD (South Africa, Nigeria, and Kenya) tend to do very well with respect to research and development.

J.T. — True. I wonder what the best ways would be to advocate for GERD increases. Regardless of where the money comes from, the key will be for people to know what funding opportunities are available to them, ideally curated in a database of sorts. I must say it is tough though because in developing countries increasing GERD inevitably means decreasing something else. But, it's a matter of reframing our mindsets that in the long term STEM can have a huge impact on the productivity of a country, much like an investment.

F.M. — That's a very good point. Many developing countries don't have the luxury to increase their spending on R&D when there are other more pressing needs. However, the UNESCO report notes that most countries are realizing the importance of science and technology for their development such that many countries are basing their long-term visions with the plan to harness science, technology, and innovation.

J.T. — Yes, a commonly held belief is that building a strong infrastructure for STEM capacity provides enormous benefits in the long term.

F.M. — So funding is one problem. It is the one problem everyone runs into when talking about developing countries. I don't think it is as simple as throwing money at the problem. I think there needs to be an appreciation of the local values. There is a need to understand what these values are so that any developments in science and technology are in line with them.

J.T. — This is such a good point. Every country has its own values. Heck, every individual has their own values. And tapping into these will be essential. I think the next key factor to consider are people!

F.M. − Exactly! What scientific questions excite them? Which technologies are important for their respective countries?

J.T. − And where do they find good mentors? Like we've discussed in the previous chapters, this seems to be the heart of the matter.

F.M. − Better yet, what is their idea of a good mentor?

J.T. − Way better!

F.M. − We can't assume what works in the developed world will apply somewhere else. But there is certainly room for all of us to learn from each other.

J.T. − And certainly the structure in the developed world arguably has its own flaws!

F.M. − True. There is room for all of us to learn from each other.

J.T. − We have a lot to learn from one another. It turns out that many countries have exchange programs where students are sent, often to the United States, to gain more training and experience. What do you think about these programs?

F.M. − I think such programs provide a unique avenue for scientific exchange. The effects are bidirectional affecting both the hosts and the visitors.

J.T. − Chile, for example, has a program called BECAS CHILE through the Ministry of Education that finances studies abroad [7]. The idea is that students will be able to bring new skills and information back to their home country to not only teach others but to thrive having gained a new experience.

F.M. − I am sure you have heard about Doctors Without Borders. Well, I recently came across Scientists Without Borders [8]. What do they do? Well, they have a focus on global health. Their online website states that "Scientists without borders works with other organizations across sectors to design, develop, and execute high-impact projects that apply creative and innovative approaches to tackling urgent global health and development challenges."

J.T. − SO COOL! And why not expand to basic science, all forms of engineering, computer science, and mathematics? I think the possibilities here seem tremendous.

F.M. − I think so! Again, I think these types of programs allow for a unique avenue for scientific exchange. Moreover, they allow for scientists to come together for a common cause: to make a difference.

You know a recent short letter from a Haitian postdoc (Dr. Gandhy Pierre-Louis) at Emory University touched on a lot of what we have been talking about [9].

He wrote:

"2016 marks the six-year anniversary of the earthquake that devastated the island nation of Haiti. The international community pledged financial support to aid in rebuilding and to orient the country towards the goal of reclaiming the title of 'Pearl of the Antilles.' However, half a decade later, the debate on how to spend the remaining funds is still ongoing. As suggested by the 2011 AAAS report [10] on Haiti, investment in science education, research, and technological innovation is necessary to bring about prosperity in Haiti. Evidence of this can be seen across several countries including Singapore, India and Brazil which until relatively recently, were in comparable economic straits. Currently, the island faces daunting threats from disease-causing microbes including cholera and chikungunya. The presence of scientists and adequate laboratories on the ground would greatly expedite diagnoses, research treatments and provide the country the means to react to future epidemics. To achieve this goal, Haiti needs to focus on building a strong science education system, which transcends didactic instructions and provides practical technical skills. Collaborations between Haitian scientists and foreign researchers are key to reaching this goal and could function synergistically. The Haiti Bioscience Initiative—a collaborative effort between science professionals in the U.S. and Haiti—is already poised to make significant contributions on that front. More of such ventures are needed to train Haiti's scientists in the appropriate technical skills, which can simultaneously lead to the support of ongoing research. With a sustained effort to increase the scientific capacity, Haiti could be transformed from an impoverished country into a self-sufficient nation that is a hub of science and innovation in the Caribbean."

J.T. — This is an incredibly well-written piece from Dr. Pierre-Louis—thanks for bringing this up. What he describes is a prime example of the simultaneous opportunities and challenges facing the developing world. Starting from nothing and building the infrastructure to do science. And also recognizing—the key point, really—that valuing STEM education can actually propel a nation forward.

F.M. — That's exactly right. I think this can be true for so many places.

J.T. — Definitely, ultimately that is the big picture. I imagine a very collaborative world in which scientists from all backgrounds and countries swap ideas and resources on research and education in the sciences. A global scientific exchange, if you will! I think that's the best chance we have to really make an impact.

F.M. − I think the Scientists Without Borders organization I mentioned earlier is getting close to this. An additional aspect of this global exchange would be in sharing "origin" stories from both young and seasoned scientists. I remember you mentioning this to me a while ago. I think hearing such stories would be very powerful. The fact of the matter is that we need to inspire young people out there to think about science and its potential. Of course, they don't have to be scientists but a scientific thinking in today's world is very useful.

J.T. − Yes!

I would love to collect stories from young scientists from all over the world. Getting to the root of what drives them, what excites them about science, what their first experiences are, and allowing them to share within a community. I think hearing these scientific origin stories would be incredibly inspiring to children and young adults, and would moreover remind the sometimes jaded, more seasoned scientists of the wonder and excitement of STEM.

F.M. − I think sharing such stories would be very powerful. It would be cool to learn about how scientists from different parts of the world got into science and other science-related careers for that matter.

J.T. − What would you ask them?!

F.M. − Simple really. What or who got you interested in science when you were young?

J.T. − I like it. I would ask them: What is one piece of advice you would give to young people interested in science? And I would want to know the best piece of advice they ever received from a mentor.

F.M. − I think we should do this. Maybe a two- to three-question survey to scientists around the world. They are busy people so I think something short should work. I am curious to see the diversity of responses.

J.T. − Consider it done! I'm game! I'm sure there will be some common threads even between different countries as well as some interesting differences.

F.M. − It would be an interesting read.

J.T. − Excited to hear the responses.

F.M. − To shift gears a little bit, the idea of a global scientific exchange got me thinking. As you well know, the current career landscape for scientists in biomedical research is a concern for many trainees. Much of the scrutiny from trainees and administrators alike focuses on ways in which the existing structure of biomedical research programs—in the United States that is—no longer serves the best interests of trainees. Many cite systemic flaws within the current academic system that could be improved to help trainees adapt to evolving academic and industrial environments.

J.T. — What kind of flaws?

F.M. — These include low rates of obtaining federal funding for research, low pay, few faculty positions, exceedingly long training periods, and well, lots of other issues [11,12]. Given this current situation, it is easy to understand how graduate and postdoctoral trainees can lose hope with the existing academic climate.

With that said, I think the current situation in the United States provides a lot of opportunities to learn for developing countries that are currently building their scientific programs. Should they use a different model? What can they do to avoid current issues facing the U.S. system? Is there a better local model they can use?

J.T. — These are all such great questions. I think the first thing to acknowledge off the bat is that there is not a uniform model for every country. We can't expect that the scientific system that works in one location will thrive or fail in another.

I'll try to approach this initially by asking why so many U.S. trainees lose hope within the academic climate (which is not easy to figure out!). Many have argued that there is an overproduction of PhDs but I don't think this is a good response.

F.M. — I think there are several answers. Yes, there are a lot of PhDs being pumped out each year into an already clogged postdoctoral system in academia. This is particularly an issue in the biosciences. Of course, you can't stop there. There are career development issues that we also need to think about. Some may think that those in PhD programs want to become scientists and that's it. Obviously, the reality is rather different. There is data all over the place showing that trainees end up selecting a wide array of careers [13–15]. And yes, being a research professor is one them.

Now, something that is not done as much is thinking more globally. Again, what about those countries that may be trying to emulate the American system? Are they likely to end up where the United States is today with respect to the biomedical research enterprise?

Can they perhaps learn from our mistakes and avoid them entirely? I don't know.

J.T. – I am honestly not sure if they will end up where the United States is today with regard to biomedical research. There is some data showing that the United States is somewhat weak with regard to STEM education [16]. So, I think the real question for other countries is not to focus so much on the academic pipeline, but to consider what skills need to be taught to young people. I think we really need to think hard about what kind of infrastructure is put in place. Here are some basic questions:

Are young people taught a core STEM curriculum?

Are they taught that STEM is widely applicable in many contexts?

Are they taught that it is okay to think outside the box and deviate from the widely perceived "academic norm?"

Are they given opportunities to experiment and FAIL?

F.M. – You are totally right. These are questions for science ministries across the world to think deeply about.

J.T. – These are really tough, fundamental questions about education. I do think that undergraduate and graduate programs in other countries have the opportunity to provide more exposure within the academic setting.

F.M. – I am fairly confident that some are already doing so in some fashion but not to the level they would like.

J.T. – Ultimately, each country will have to define, within their own unique cultural and societal experiences, what their STEM pipeline will look like from school age children, high school, college, graduate school, and ultimately academia and beyond.

F.M. – That's true. The STEM pipeline definitions have to be intrinsic to their own needs. Countries have to think about what they value most for their own future.

J.T. – I'm trying to think about a good framework for countries to consider. Building a STEM pipeline certainly seems like a daunting task.

F.M. – It is a daunting task but like any task, you can break it down to its components. It is akin to constructing a building. You need a good foundation which you then build onto. The same goes for STEM. There needs to an intensive focus on the basics in STEM education and then work your way up.

J.T. — So let's break this down from the bottom up! And just to reiterate that in the developing world, the specifics of building a STEM pipeline will vary but are critical to consider.

F.M. — So I personally think that governments need to come up with a vision statement that all key stakeholders across sectors would be able to champion behind.

J.T. — Easier said than done, but I absolutely agree.

F.M. — It is important that this task is difficult. It needs to be difficult. It needs everyone to be at the table. It may sound naive but it is critical that there is a vision that everyone sees. Think about the vision of traveling to the moon that was set by President John F. Kennedy.

J.T. — Government buy-in seems essential otherwise one is stuck in gridlock.

F.M. — True. But the usual argument for STEM education that most governments agree on is the idea that STEM-oriented economies perform strongly on a number of economic indicators. Having this buy-in provides a framework into thinking deeply about a country's own STEM vision.

J.T. — Frankly, this actually may be the most challenging step of all. Other critical components to this foundational structure include personnel, particularly well-trained teachers, and then physical infrastructure—buildings for schools and universities. Perhaps the physical buildings are less important these days with technology, but the concept of academic institutions whether at the primary school or university level is important.

F.M. — You are absolutely right especially on having well-trained teachers. Of course we need to make sure students are able to eat, get to school safely, and have access to the best teachers and school facilities.

J.T. — Spot on.

F.M. — It is a comprehensive approach and that's why I think there needs to be a powerful vision that is able to galvanize a whole range of people and agencies toward achieving the vision.

J.T. — Okay, so let's say arguably the hardest part has been achieved—a feat in and of itself! Then we fill in the details, mind you, important details. I like thinking about this in terms of child development. Does preschool exist? How do children obtain basic language and communication skills? In what context do they learn to interact with other children and with adults?

F.M. — I don't see why preschool wouldn't exist.

J.T. — I'm not sure it exists in every country!

F.M. — I guess it depends on your definition of preschool. But I don't know if that's the right question.

J.T. — Take two. Perhaps the better question is once the infrastructure is in place (which I also forgot to mention will certainly require maintenance), how do you get young people to school?

Not only do they need a physical means of transportation, but they must also have enough social stability at home.

F.M. — I will admit that I am optimistic that once the vision is set, these things will naturally get sorted out. They have to or else the vision wouldn't work.

J.T. — I disagree. The tough part for me is that when I start thinking about all of the components necessary to get a kid to school, my head starts spinning! Are food and water available at home? Do their parents value education or would prefer them to work at home? Do they have any family members who are ill? Does the school even have electricity, running water, papers, pencils, and books?

F.M. — Of course your head will start spinning when you think about all the individual components in a vacuum. But add a grounding vision and see what happens. What I am suggesting is for a country to employ foundational thinking about its priorities in education, health, and all the other important things.

J.T. — I think a foundational vision is key but we can't just assume that the details will simply fall into place. There are a lot of moving parts to consider. The last really key component I think about is the environment in which kids learn. Is learning punitive where you are punished for getting the wrong answer or is it constructive where you learn from your mistakes and are encouraged to experiment? I think it goes without saying what my preference would be.

F.M. — You need to be able to learn from mistakes. I think that's where "real" learning takes place. Again, this is something that needs to occur at a foundational level. It must start early for its impact to be truly felt in the long run.

J.T. — Is there anything else you would include? This basic infrastructure is only a starting point for countries who are building a STEM pipeline. But ultimately, you are right, that the overarching vision must permeate everything.

F.M. — I would actually argue that the same principle applies for those countries with advanced STEM pipelines. It is easy to make assumptions that the foundation is in great shape. It doesn't hurt to occasionally test those assumptions.

J.T. – And actually, when you break it down, the same principle applies to pretty much anything that you are doing! If you're slogging through a PhD program, you always have to keep your vision in the back of your mind.

F.M. – Interesting point. How do you think a vision trickles down to the individual?

J.T. – Oh wow. I have to think about this one.

F.M. – It isn't easy but let me ask a related question: Can a vision inspire a story? Of course I am talking about science here.

J.T. – Hmm, what do you mean by a vision inspiring a story?

F.M. – Let's go back to my earlier statement regarding President John F. Kennedy's speech in 1962. He said the following: "We choose to go to the moon in this decade and do the other things, not because they are easy, but because they are hard, because that goal will serve to organize and measure the best of our energies and skills." Something grandiose like this is truly inspiring. I truly believe that such visions have the capacity to trickle down to the individual. It takes the right leader to deliver the vision.

J.T. – The next natural question (which is fun to think about) is: How do WE inspire the next generation of scientists?

F.M. – That's a good question. It is something I shared on the 2014 Nextgen VOICES survey hosted by Science Magazine [17]. For me, I would focus on mentoring and giving talks to high- and middle-school students. I strongly believe that strengthening the quality of mentorship within the STEM pipeline is crucial if we are to strengthen and diversify the STEM pipeline. I was lucky to have had the exposure, support, and guidance of my parents, science teachers in high school, professors in college and graduate school, and many informal mentors. Many students do not have this kind of support.

If it were up to me, I would also focus on advocating for the strengthening of mentorship channels by encouraging universities to provide incentives for their staff and professors to spend more of their time mentoring and improving their mentorship skills. In doing so, it may be possible to inspire more people to get into the many parts of science (academic scientists and teachers, science administrators, policy analysts, and so forth). Science has so many sides.

J.T. – I agree completely. For me, the key is capturing the attention and fascination of students when they are in grade school.

Ultimately, mentorship is a theme that we are seeing pop up throughout our discussions. I have spent a lot of personal time in one-on-one recruitment to really help undergraduates, medical students, and graduate students make informed decisions about their next steps forward in their scientific lives!

I think we have to always give back, no matter what stage we are at. And you are so right. Science has an incredible number of facets. Everyone has something incredible to contribute and share.

F.M. − Yes we can! I think inspiring the next generation deserves its own chapter.

J.T. − Absolutely! What are tangible ways in which we can give back and, in the process, pay it forward? We'll discuss this and so much more in Chapter 4, Inspiring the Next Generation!

REFERENCES

[1] List of organizations engaged in STEM education across Africa, < https://en.wikipedia. org/wiki/List_of_organizations_engaged_in_STEM_education_across_Africa?wteswitched = 1> [accessed 12.03.17].

[2] UNESCO Science Report 2010, < http://unesdoc.unesco.org/images/0018/001899/189958e. pdf/ > [accessed 01.11.16].

[3] Teaching and Research in Natural Sciences for Development in Africa, < http:// Trendinafrica.org > [accessed 11.09.16].

[4] Muindi F, Guha M. Global fund needed for STEM education. Nature 2014;506(7849):434.

[5] Global Fund to Fight AIDS, Tuberculosis and Malaria. < http://www.theglobalfund.org/en/ > [accessed 16.02.17].

[6] UNESCO Science Report Towards 2030, < http://en.unesco.org/node/252168/ > [accessed 21.01.17].

[7] BECAS   CHILE,   < http://portales.mineduc.cl/index.php?id_portal = 60/ >   [accessed 21.01.17].

[8] Scientists Without Borders, < http://www.nyas.org/WhatWeDo/ScientistsWithoutBorders. aspx/ > [accessed 21.01.17].

[9] Pierre-Louis G. Education: make science a priority in Haiti, < http://www.stemadvocacy. org/single-post/2016/1/20/Education-Make-science-a-priority-in-Haiti/ > [accessed 20.01.16].

[10] Machlis G, Colon J, McKendry J. Science for Haiti: a report on advancing Haitian Science and Science Education capacity. AAAS 2011, < https://www.aaas.org/sites/default/files/ migrate/uploads/haiti_report_2011.pdf/ > [accessed 21.01.17].

[11] Schillebeeckx M, Maricque B, Lewis C. The missing piece to changing the university culture. Nat Biotechnol 2013;31:938−41.

[12] Alberts B, Kirschner MW, Tilghman S, Varmus H. Rescuing US biomedical research from its systemic flaws. PNAS 2014;111(16):5773−7.

[13] Gibbs KDJ, McGready J, Bennett JC, Griffin K. PLoS ONE 2014;9:e114736.

[14] Fuhrmann CN, Halmem DG, O'Sullivan PS, Lindstaedt B. CBE Life Sci Educ 2011;10:239−49.

[15] Mason MA, Goulden M, Frasch K. Why graduate students reject the fast track. Academe 2009;95:11–16.

[16] DeSilver D. U.S. students improving – slowly – in math and science, but still lagging internationally, < http://www.pewresearch.org/fact-tank/2015/02/02/u-s-students-improving-slowly-in-math-and-science-but-still-lagging-internationally/ > [accessed 06.01.16].

[17] NextGEN Results 2014, < http://science.sciencemag.org/content/344/6179/34/suppl/DC1/ > [accessed 20.01.16].

CHAPTER 4

Inspiring the Next Generation

J.T. — So, what do you think is the difference between giving back and paying it forward?

F.M. — That's an interesting question. There is a lot of talk about giving back. I understand this and I think it is important—to some extent—to do so. However, I think it is just as important to pay it forward. You see, giving back typically focuses on one's own community from where he or she came from. It hints, to some level, that you are paying BACK something to the community that helped you get to where you are. Of course, it is important to give back. But consider the following: How about individuals or communities that didn't help you directly? How about communities that are far away, communities that look nothing like you, and who know nothing about you? Are they not as important? For me, this is what paying it forward means. You are passing on what your own community gave you to other people outside that community. This is actually an old concept. Personally, I do both. Both are important and should be encouraged. Perhaps they mean the same thing to some people. That's fine too.

J.T. — I think this is an incredibly important distinction, and I like this concept that you've described a lot. You're absolutely right. For me, I imagine taking everything I've learned in the past—all the struggles, successes, failures—I carry all of these lessons with me. The word that keeps coming to mind is openness. What are examples of how you have done this? I think it can take many forms.

F.M. — "If you had five extra hours per week to devote to advocacy for science, how would you use that time?" This was a question that was asked as part of NextGen VOICES series in the April 2014 issue of Science Magazine [1]. I wrote that "I would devote it to mentoring and giving talks to high school and middle school minority students." I went on to write that "strengthening the quality of mentorship within the STEM pipeline is crucial if we are to diversify

Journeys in Science. DOI: http://dx.doi.org/10.1016/B978-0-12-813090-2.00004-X

the STEM workforce." As I reflect back on this statement, I am happy to have had the opportunity to do exactly what I wrote. Back in 2015, I was invited to talk to middle school students at Mary E. Curley Middle School in Boston. For 45 minutes, students asked a whole load of questions about my research interests, and more importantly, they asked about my journey in science. I was humbled to stand in front of that classroom and share my story. My take home message for them was to never give up. I said: "Things will get difficult. You will likely find yourself not understanding some concepts in class. You may find your internal dialogue filled with negative comments. Don't get frustrated. Ask for help, but whatever you do, don't give up so easily." I still think that when time permits, such activities should be encouraged for graduate students, postdoctoral fellows, and even professors. Sure, we are all so busy and there is little time to do such things. However, I think there is always time to pay it forward. Imagine if more of us did outreach events in middle schools and high schools! It would make a noticeable difference.

J.T. – That's awesome—I'm sure the kids had such a great time asking you questions and meeting a real-life scientist! When I was a Graduate Public Service fellow, I spoke to some elementary school students in East Palo Alto. They asked such insightful questions, and I think I learned just as much, if not more, than they learned from me. There is an incredibly contagious excitement that students have about science. You can just see it on their faces—the curiosity and the wheels just churning away in their heads. I really think middle and high school students are thirsting for more cool STEM!

F.M. – I totally agree. I have a hunch that kids in developing countries have an immense thirst for STEM.

J.T. – Definitely, definitely. There are so many ways to share with your community. Do you remember when we were both on the Graduate Student Council?

F.M. – Ow yes. It was my first taste of student government. It provided a powerful avenue to engage the wider graduate community.

J.T. – For me, there is something to be said about becoming engaged in your community (whatever form that might be) in a more active way. I have always seen value in thinking about and, more importantly, in finding ways to improve the communities that I am part of.

F.M. – The problem, of course, is time.

J.T. – So how do you manage time?! That's the ultimate question.

F.M. – Research always comes first, of course. It has to be first. Once the research is on solid ground, it becomes easier to allocate time to outreach activities. It then becomes a delicate balance to ensure the research doesn't fall too far behind.

J.T. – Absolutely agreed. I think a lot of it requires diligent time management. Even outside of STEM, time management is essential. I find myself doing a lot of planning and scheduling in advance to try to set myself up well. I think there is a lot of managing expectations, particularly for students. If outreach is something that is important to you, it is worthwhile to express this to one's advisor or mentor.

I also hear from a lot of people that paying it forward seems like a huge time commitment. Do you have some tangible examples of simple ways to pay it forward?

F.M. – I was fortunate enough in graduate school to dedicate quite a bit of time to outreach. With the help of some friends, I founded a small group called Run4education. What did we do? Well, we literally ran races to raise funds for local schools in the East Palo Alto area. That endeavor taught me a lot.

J.T. – I remember! You are a very fast runner by the way...

F.M. – I used to be. Now I take it easy. Anyway, the time commitment there was quite considerable. I don't recommend starting a whole organization. Of course, some may have the bandwidth to do so. For me, I managed to find myself in front of classrooms talking to middle school and high school students quite often. The time commitment there wasn't so bad. One has to just show up at a classroom and share his/her passion for science and research!

J.T. – I think this is a very important point, by the way, because outreach does not have to be a huge time commitment and certainly should never feel like a burden. I always try to emphasize to people that there are actually really simple ways to make a big difference. Even in day-to-day life, sometimes a friend of a friend will connect me with someone who is interested in neuroscience, pediatrics, or science in general. Having a quick chat via email or phone can make such a difference.

F.M. – And that's all it takes really. So, anyway, the Run4education group didn't last long because it relied too heavily on the running! But the experiment taught me something important about myself: my passion for science education was undeniable. In particular, I was moved by the potential of science to change the world. I am still driven by this. Anyway, I was energized with the idea of inspiring the next generation of science policy makers, academicians, doctors, and so forth.

You know—as I think about it now—my outreach efforts helped me realize that I was part of something much bigger than myself. I think this is something very important to realize. I like what Dr. Carl Sagan said: "Science is a way of thinking much more than it is a body of knowledge."

J.T. – Awesome, awesome, so AWESOME!

> Sharing passion is one of my favorite things to do. There is this feeling when you are in the moment talking to somebody about something that you really enjoy, where you are smiling, gesticulating wildly, and talking really quickly. That's a great feeling. And I think it really shows. The excitement and enthusiasm cannot be contained—it's infectious!

F.M. – It really is. I am always looking for opportunities to share my excitement and passion for science with others. Anyway, back to your question. I think one can get creative in finding the time to pay it forward. You really don't have to start a whole organization. Just do the little things. That's all it takes.

J.T. – One way I remind myself of this is just to think about all the people who paid it forward to me. It was each coffee chat, a quick conversation in the hallway, a short email—all of these things have impacted me in incredibly positive and productive ways!

F.M. – Very true.

J.T. – And now, technology allows this to happen at a much broader scale through Twitter, Facebook, and various blogs.

F.M. – They act as multipliers.

J.T. – It's sort of crazy if you think about it. You could easily talk about mentorship and science with someone you have never met, across the globe! The possibilities are really endless...

F.M. – Nothing beats face-to-face though!

J.T. – Yup! Skype or FaceTime hah!

F.M. – I have to mention the Big Brothers Big Sisters program in Boston. It is a great program to get involved in. The commitment is serious (minimum 1 year, I believe). They place high importance on mentorship. Of course, there are several other programs that may be better suited for different people. One should just look around. There are tons of opportunities.

J.T. – Great program, I have a dear friend in San Francisco who is very involved in Big Sisters. What sorts of things have you and your little brother done?

F.M. – Gosh, quite a bit. I can't believe it's been just over 2 years now! I have realized that the most important thing for him was for me to just be there. To show that I cared. I have taken my responsibility very seriously. We hang out on average twice every month. We have explored museums, visited my research laboratory at MIT, discussed my research, gone for walks, seen shows, rode our bikes, and lots of other things. He recently turned 13 years old. Can you believe it? Anyway, with his help, I wrote a short article describing some of the activities we did over the last 2 years [2]. It was great to hear his feedback on the article!

J.T. – What has been the most important aspect of being a solid big brother?

F.M. – I think showing up (yes showing up), being an active listener, and demonstrating genuine care for his development have all been important. During my interactions, I have realized how easy it is to make a difference in someone's life at that stage. You listen and make positive suggestions for them to consider. You essentially make suggestions for small course corrections here and there. The key is to be consistent. As such, I make the effort to communicate with him as often as I can. The time commitment is worth it. Someone did it for me. I will do it for him. I only hope he will do it for someone else in the near future.

J.T. – I'm sure he will! I'm glad you said that you realized how easy it is to make a difference in a child's life. That's one of my favorite things about Pediatrics!

Wouldn't it be nice to have an easily accessible (nonexhaustive, of course) list of advocacy and community opportunities for paying it forward? And I'm just riffing here, but maybe even a forum that would facilitate connections between people, allowing them to discuss their experiences with these activities. A crazy idea, I know, but it could be so fun and powerful!

F.M. – The question is whether something already exists. I would recommend people look into the Connectory (https://theconnectory.org/) and also the STEM Connector (http://stemconnector.org/). Both provide impressive online databases that should certainly be explored.

J.T. – Yup, will have to look into that!

F.M. – As I mentioned earlier, there are several programs for those interested in getting started. In fact, I had an insightful discussion with someone with a great deal of knowledge about mentoring programs in the Boston area.

J.T. – Wow, very cool—would love to hear more!

F.M. – Mass Mentoring was something he told me about.

J.T. – What's Mass Mentoring?

F.M. – Founded in 1992, their website states that they are focused on empowering youth–adult relationships across Massachusetts. They serve more than 250 mentoring and youth development programs statewide supporting more than 33,000 youth in mentoring relationships [3]. It is quite an impressive list of activities they are involved in.

J.T. – That's incredible. I'm thinking about the power of all those mentoring relationships combined.

F.M. – Very true. Anyway, our discussion focused a lot on the importance of keeping in mind the diversity of student's backgrounds. This is something I hadn't really thought deeply about. A student's background greatly affects their ability to get into and stay in the educational pipeline. Of course that includes STEM.

J.T. – Ah, I love this topic. I think traditionally people think of diversity as an issue of race alone. But, it is so much more than that.

F.M. – Way more. It covers a large range of areas. Socioeconomic, access to health services, teacher quality, access to resources, mentorship, family stability... care to add?

J.T. – Absolutely—educational background, formative life experiences, sexual orientation, religion, spirituality, overcoming major obstacles, hobbies, and more.

F.M. – But you know, diversity at the level of race is important to mention.

J.T. – There's no denying it.

F.M. – Let's face it. There are very few racial minorities in the sciences.

J.T. – Why do you think that is? I was the first person in my family to get an advanced doctoral degree.

F.M. – That's amazing! Your parents must be so proud!

J.T. – I think they are, but it hasn't always been the easiest road.

F.M. – So, I personally don't think there is ONE thing alone that you can point to and say this is "THE" reason that there are so few racial minorities. It is a combination of factors and I think there has been a great deal of discussion and publications about this topic.

J.T. – Let's break it down.

F.M. – We could! But, I think what may be more useful is to share tips on how young people can approach the challenge of finding a lack of minorities and women in science.

J.T. – Let's do it!

F.M. – It can be a real challenge when you are unable to find mentors who look like you. To some extent, I think it is absolutely important to find mentors who look like you. But you shouldn't stop there. There are great mentors everywhere and they come in all shapes, sizes, and colors. What has helped me throughout my journey is to find good mentors who are willing to take the time to guide me and reach out to me when things are not going great. I have been lucky to find good mentors along the way.

J.T. – I've realized honestly that there are not that many women, let alone Asian American women, who are physician-scientists in academia. So the reality is that it is difficult, I think, to find mentors who look like you.

F.M. – I agree. It is a numbers game really. There simply aren't that many minority scientists to begin with. So, naturally it is more difficult to find mentors who look like you, especially in science. But let me ask this question: Is it important that we find people who look like us? Are we not automatically eliminating a lot of potentially good mentors when we do this?

J.T. – A truly good mentor is color blind—they should support you fully no matter what you look like, where you come from, or what experiences you have had in your life. To be perfectly frank, I have never had a physician-scientist mentor who is an Asian American woman. I consider myself very lucky to have many mentors who are outstanding and have supported me every step of the way. In some ways, I think it would just be comforting to have a mentor who had a similar background to mine. It would be reassuring to know that someone who looks like me can make it. Does that make sense?

F.M. – I understand that. This absolutely makes sense. It is indeed true that having such a mentor—one that comes from a similar background—can certainly help you traverse the world of science and beyond. So one can say it is necessary. But is it sufficient?

I personally think it is about having a diversity of mentors who come from multiple backgrounds, one of which may include your own. This way you are able to learn many ways to succeed in science because success in science is about diversity. Diversity in your mentors. Diversity in your research. Diversity in your thinking. Diversity in pretty much everything. Look at nature! It screams diversity. I think this message is key for young minorities and women who are entering science.

What do you think?

J.T. − Diversity in all things, couldn't have put it better myself. Although, let me spin this question back at you. Do you think there are young people who are discouraged by not having a mentor who looks like them or who comes from a similar background? I worry that incredible people decide not to pursue science because they feel that they just don't belong.

F.M. − I absolutely think there are young people who are discouraged by not seeing people who look like them in science. It is one of the main reasons I try to go out and talk to young people about my experience in science. They need to see that someone who looks like them can "make it," if you will. My message is always that my success came from a collection of people who took time to mentor and help me along the way. They were women. They were men. They were Black, White, Asian, old, young, and everybody else in between. In my talks, I try to leave young people with a message of hope that they can indeed succeed as well with a little help from a multitude of people, some of whom may not look like them.

J.T. − I do think this is the key, but do you ever feel like you have more responsibility as an African American in science? I feel quite a bit of self-imposed pressure as a woman in science.

F.M. − Certainly. It is an honor to have such a responsibility. As I mentioned earlier, it is about paying it forward. I think there are multiple ways to do so both inside and outside of science. I think part of the pressure comes from the feeling that one can no longer be that role model if he or she leaves bench science. Again, I think there is plenty of room for all of us to play a role in getting more minority young people into science as researchers and also using science to make an impact in policy, industry, and education, just to name a few. But wait, can you explain a little about this self-imposed pressure you feel as a woman in science?

J.T. − Honestly, I was blind to all of this until medical school. I've always been motivated by the science, by the questions, and by an inner drive to just do the best that I can. There is a tremendous satisfaction that comes, regardless of who you are, from knowing that you have done the best job that you can. I can remember the moment that I felt that self-imposed pressure. I was on the MD-PhD admissions committee and during a meeting I looked up from my stack of papers, and I realized I was the only woman in the room. I was the one woman in a room full of men! It hit me like a train. I had so many questions. Why were there no women at this meeting? Was it that no one had been invited? Was it that they had been invited, but could not make it? What implications did this have for admitting more women to the program? Were we biasing our admissions process without even knowing it? So, I think this is the setting where I developed self-imposed pressure. I feel an enormous responsibility to support other women and really encourage and nourish their scientific endeavors. The crux of it all circles back to what you said—the bottom line is more diversity of thought.

F.M. – It really does circle back to diversity.

J.T. – I'm curious if you ever feel this pressure. I occasionally have this feeling that all eyes are on me like a hawk. I wonder if people are just waiting for me to fail and watching my every move. So, partially for this reason, I do think I work excessively hard. And don't get me wrong, I'm not saying this is right by any means. But, I do feel that I have a whole lot to prove perhaps not even to myself, but to other people.

F.M. – This is interesting. It's internal for me. I have found ways to handle it though. It is key, especially for minority students, to find good mentors to be by their side to tell them "you are on the right track; don't give up, you will find a way," and many other inspiring comments. As you mentioned, there is an outside pressure one can feel. But there is also an internal pressure that can wreak havoc if left unchecked.

J.T. – The internal pressure is real. I remember it was tough even in middle school and high school. My parents were born in another country and moved here as immigrants. And while they spoke English and were both engineers, I remember always being slightly jealous of my classmates whose parents would edit their essays and homework. There was no way my parents would be able to do that for me. So I had to buckle down and figure it out for myself.

F.M. – You took the initiative. That's also important throughout life.

J.T. – Yes, but again, I worry that others will simply shut down, become complacent, and give up.

F.M. – So how can one avoid the complacency and shutting down?

J.T. – At the core is building resiliency.

F. M. – Which I think comes back to diversity. Resilience through diversity in multiple dimensions as we talked about before.

J. T. – So the big question is how do all of these young people find a diverse set of mentors? It seems like a daunting task!

F. M. – It is. I think part of the trick is to start early. It takes time, so don't wait until the last minute. Finding the right set of mentors at a particular moment in your life takes time. You are likely to get it wrong. But that's okay. So start early and talk to a lot of people. But listen very carefully and talk selectively. It's challenging especially when you are young.

J. T. – We have to teach children at a very early age to seek out mentorship. It's not clear to me that this is necessarily taught even at the middle school or high school level!

F. M. – It's tough.

J. T. – Especially since young people come from such varying backgrounds—it is really tough to standardize this stuff.

F. M. – With these differences in mind, it is then reasonable to expect that these students need mentors from a variety of backgrounds. They need to hear stories from a wide range of people who have successfully transitioned through and faced the many ups and downs of the educational pipeline. I think stories have the power to make a difference.

J. T. – Yes, this was the impetus for gathering scientific origin stories from people. We heard amazing stories from an incredibly diverse group of people ranging from students to CEOs to professors and more!

F. M. – Yeah, the project was inspired during our discussions right here in this book. Basically we asked a diverse group of people the following three questions:

1. **What or who got you interested in science when you were young?**
2. **What is one piece of advice you would give to young people interested in science?**
3. **What is the best piece of advice you ever received from a mentor?**

And then the fun began. It was very inspiring to read their responses. In fact, the diversity of their responses was rather insightful.

J. T. – Reading their responses was by far the best part!

Let's start with Dr. Nathan Vanderford who is an Assistant Professor at the University of Kentucky. He wrote the following:

What or who got you interested in science when you were young?

"I am a first generation college graduate from a small, rural town in Tennessee. Despite limited educational resources in the area, I developed a love for math in high school, and after taking courses in basic biology, chemistry, and physics, I developed a deep desire to learn everything I possibly could about how biological systems work at a molecular level. This desire gave me tremendous motivation to leave my local community in order to have access to the level of science education that would fulfill my thirst for learning about biology. So, what or who got me interested in science? Well, in looking back on this, I had a couple of good teachers in high school who helped instill in me an interest in science, but without meaning this in a narcissistic way, I truly believe that my self-motivation ultimately drove my interest in science when I was young."

What is the best piece of advice you ever received from a mentor?

"Sometimes actions are more powerful than words. The greatest support I have received from mentors has been through their actions. A great mentor is someone that will champion your dreams and causes through tangible actions. Find these people and you will be successful. Once you are in a position to do so, be a genuine champion to as many mentees as possible."

What is one piece of advice you would give to young people interested in science?

"Dream BIG dreams and pursue those dreams with everything you have. Dream about having an impact on the world by being a behavioral scientist, biomedical researcher, a physician-scientist, a chemical engineer, a bioinformatician, etc. Find individuals (aka, mentors) that will provide encouragement and resources to help you achieve your dreams. Never let a roadblock, including negative people, stand in the way of your dreams. Once you're in a position to do so, do everything in your power to help the next generation achieve their dreams."

F.M. – I think the first question is tough for most people. What do you think?

J.T. – Definitely, it seems easy to forget the moments that sparked our interest in science and technology. I wonder why that is?

F.M. – It is those early moments that most of us forget. The funny thing is that it is these early moments that people really need to hear, especially the young generation.

J.T. – I wonder if the farther in time we get from those moments, the more likely we are to forget. I've found it incredibly motivating to push myself to remember those moments. The aha! moment, the light bulb moment, the "wow this is insanely awesome" moment, the "could I actually get paid to have this much fun?!" moment.

F.M. – I guess this is tough when you are young. I think middle school and/or high school teachers play a huge role for most people. It was certainly the case for Dr. Vanderford. I think we have talked extensively about how important teachers are. They really do hold a lot of influence.

J.T. – Very true. Sure, the science is interesting on its own. But without knowing anything about the natural world or science or engineering, the teachers are the ones who can creatively show young people why science is cool.

F.M. – Something important that Dr. Vanderford mentioned is self-motivation. This is tricky though. I am not sure you can teach someone motivation. All you can do is encourage them.

J.T. – That intrinsic motivation is interesting. I noticed a trend in responses that a lot of people talked about this curiosity of finding new knowledge and discovering how things in the world work. Others noted an intrinsic interest in helping and bettering the status quo.

F.M. – To make a difference in some way.

J.T. – I really like Dr. Vanderford's description of a solid mentor when he writes, *"A great mentor is someone that will champion your dreams and causes through tangible actions."* This is something we should all aspire to become!

F.M. – I think mentorship is crucial. I think some of us wait a little too long to be in a position to mentor. The funny thing is that many of us are in a good position to mentor. You don't have to have tenure to mentor. You can start small and build your mentorship skills. I really do think that mentorship should be a class or something in middle school and high school. They should even make it mandatory in college, and you know, even graduate school. Why not?

J.T. – I know I could certainly learn a lot from young people! I am really drawn to this attitude of sharing and collaborating as a way of teaching. Everyone has their own expertise and their own experience that they can lend to others. One of the reasons I love science is that you are never "finished." Learning new information leads to more questions!

F.M. – That's very true. On a different note entirely, it really is about the questions! In school, we learn how to answer questions.

We should be learning how to ask good questions. Anyway, that was a side thought. Back to mentorship.

J.T. – The concept of mentoring as actions really resonates with me. It's a nice spin on the definition of mentorship.

F.M. – Getting back to origin stories, I think Dr. Armon Sharei (CEO of SQZ Biotech) gives good insight into those early experiences that got him into science.

What or who got you interested in science when you were young?

"I actually don't know. I feel it was just a more instinctual thing in me that would get excited about it. Or maybe it was a combination of things. e.g., a) My uncle would get me model rockets that I used to love flying and was fascinated by how their engine cartridges worked to produce such a kick. b) Of course I found Bill Nye the science guy interesting even though I only ever saw a handful of episodes because I lived in Iran. c) My dad was good at hyping up various scientific concepts that always interested me. At the end of the day, I think the main reason I committed myself to science is because I wanted to make the world a better place. And yes you can do that through entering government or working with charities etc., but if you make an advance in science or medicine that improvement to human life is permanent, treaties can be broken, government policies can change but no one can ever take back antibiotics or the airplane. So the irreversible impact of science is ultimately what made me take the plunge. Age wise that stuff has been going on for as long as I can remember so minimum age of 5 I would say. The model rockets started before the age of 10 I think (my uncle didn't pay attention to age limits on the box. . .pretty sure that stuff was meant for adults only). My dad would try every possible topic, nuclear energy/bombs, computers, medicine, physics etc. He is more of a computer science guy by background so he didn't know much chemistry or biology."

What is one piece of advice you would give to young people interested in science?

"Go out and learn more about it because there is a fascinating world of it! Perhaps get some cool science related hobbies. E.g. flying model rockets and model planes."

What is the best piece of advice you ever received from a mentor?

"Follow your heart. Which I interpret as always make the decision you can get behind as opposed to just doing what people tell you, even if they have much more authority/experience over the matter. . .doesn't mean they are right."

J.T. – It looks like you and Dr. Sharei both had a fascination with airplanes! I really like what he wrote to us about the irreversible impact of science—making the world a better place in a permanent way.

F.M. – Airplanes are truly fascinating. I still think so. But yes, the impact of science in the world is certainly evident. But we have to be careful here. We shouldn't give science so much power. Science is important? Yes! But I see it as one path (among others) that opens up the mind in unique ways. Is science the only thing young people should strive toward or think about? Certainly not. Airplanes were fascinating to me because of their magic. They flew! And in this amazement, I wondered how it all happened.

J.T. – The idea of flying still boggles my mind every single time I take a flight! I don't think that feeling will ever go away.

Fanuel, you make a really good point. Science is certainly not the end all, be all. The arts, humanities, and social sciences are equally important. And even more powerful is the combination of these disciplines.

F.M. – Exactly. In my humble opinion, the fusion of the arts and sciences is where the magic happens. This is something that should be encouraged. Actually there are some efforts to this end. Ever heard of STEAM? Science, Technology, Arts, and Mathematics. There is an initiative known as STEM to STEAM that is doing exactly this [4]. The idea is simple: *STEM + Art = STEAM*.

J.T. – THIS IS SO COOL. Going back to the concept of collaboration, there is so much opportunity for scientists to learn from other disciplines. This is how we get creative in our investigative approaches and think outside of the box.

F.M. – In order to be innovative, I think you need to collaborate. It is almost necessary in today's world.

J.T. – Totally agreed. I really appreciate the best piece of advice that Dr. Sharei shared with us. This goes along with the idea that even young people have a lot to teach very experienced people. It's important to stick to your gut.

F.M. – It's a delicate balance for sure.

J.T. – I love hearing about how people got interested in science. These are such important moments in people's lives!

F.M. – Very true. I really like what Dr. Matthew Carter wrote on how he got into science.

J.T. – His inspiration came from outer space! He wrote:

"I loved Cosmos by Carl Sagan, because he asked questions that seemed impossible to answer, then showed how people throughout human

history had ingeniously figured out ways to answer them. He also made
science seem like one of the most worthwhile pursuits one could do as a
human being.

I just really liked planets and learning that there were different worlds
that could be very different from the earth. That was the entry point to
finding Cosmos and to learning more about nature in general. Seeing
pictures from the surface of Mars and Venus really made me realize how
limited my tiny view of the world was."

F.M. — I think I have already mentioned Carl Sagan before. A couple of times actually.

J.T. — Haha, yup, you have!

F.M. — It's interesting that I came across Carl Sagan at a much later stage in life. In fact, it was not until my postdoctoral studies at MIT that I came across the show Cosmos. I would certainly have been inspired had I come across Carl Sagan's words when I was younger. I would agree with Dr. Carter that Carl Sagan made science seem like something worthwhile to pursue. His words are truly inspirational even today.

J.T. — Carl Sagan inspired (and continues to inspire) millions of people worldwide!

One of my favorite memories from elementary school was a class field trip to the planetarium. I remember staring up in the pitch black room, completely in astonishment and absolute wonder. Did you ever go to a planetarium? It's quite an experience.

F.M. — Planetariums are awesome. Every kid should go to one. The one in San Francisco is pretty good.

J.T. — Every adult should go too!

F.M. — Definitely. Actually, adults MUST go.

J.T. — You know, the other thing I like about Dr. Carter's origin story is the idea of asking impossible questions. And even more insightful is the concept of being able to answer all questions in amazingly creative ways!

I actually had an experiential moment during my PhD when I was preparing for my qualifying exam. I was reading some old papers from the 1970s to 1980s by Michael Brown and Joseph Goldstein who together won the Nobel Prize for their understanding of cholesterol metabolism. The papers are so simple, so elegant, and so creative in answering fundamental questions. I was astonished by how clever they were without the technology that we have today. I guess this also goes to show that these "origin stories" can occur more than once. Sure they can happen when you are a young child, but they also recur later in life! Moments that remind us why science is so cool.

F. M. – In fact, Dr. Tom Baden touches on this in a nice way. He writes the following in response to what got him interested in science:

"I guess I was always drawn to things that are logical, and tickled by the fact that there are a lot of things in nature we just don't understand yet. The journey to the edge of current knowledge is often a rather short one, and what lies beneath will almost never disappoint if you bother to look."

J. T. – Wow, I really could not have said it better. The fun part is that the edge of current knowledge is always shifting. Some people may find this irritating—that there's no end in sight. But, that's really the fun part, don't you think?

F. M. – I certainly think so. How do we teach this to kids? How do we get kids to not worry so much about the answers but get them excited about the questions? It is essential for kids to learn how to ask important questions. That's what makes science so exciting. New questions challenge what we think we know to be true.

J. T. – Great point—this is such a tough one. On the one hand, there is so much focus in education on getting the right answers and scoring points. On the other hand, we also want kids to learn the fundamentals. I don't think these two things are mutually exclusive, but how to accomplish them both is an interesting challenge.

Questions not only shift our general knowledge. They drive you to think in systematic ways. You think to yourself: "If one does X, what happens? What if one does Y? What if you tweak Z in this tiny way? What about combining a bit of X and a bit more of Y?" And then, things get even more powerful when multiple people come together.

F. M. – Dr. Matthew Carter says it really nicely:

"What it means to 'do science' changes tremendously throughout a person's life. In middle school and high school, science is about learning the basics. In the early years of college, science is about rigorously learning the fundamentals. In the latter years of college, science is about reading the literature and learning how science is pursued in modern laboratories. In grad school and as a postdoc, science is about producing original research and generating knowledge yourself. And as a PI, science is about managing several other scientists and asking much larger conceptual questions. Each stage is completely different from the others, and it is possible to not enjoy science in high school but love it in grad school."

J. T. – I love this answer. Let me piggyback off of this. I actually think it is possible to enjoy science and engineering in all stages, but to enjoy different things about each stage, if that makes sense? What do you think?

In many ways, you gain more self-awareness as you move through the stages, so you are able to grasp what your strengths and weaknesses are. This enables you to play to your strengths while working on weaknesses. I also think an important part of Dr. Carter's answer is that many people don't realize how different the stages really are.

F.M. – It totally makes sense. Each stage is different. Each stage presents a new set of challenges and a new set of questions. Each stage presents new opportunities for growth.

J.T. – So, the best part, and we've discussed this before, is how do the stages combine to form a story? What is your trajectory, what is your arc? How do you take all of the stages into account?

I wish I could draw this out for you, but this is what I mean (*x*-axis is time):

```
Stage 1     ---------
Stage 2     -----------------
Stage 3     ----------------------------
Stage 4     ---------------------------------------
Stage 5     ---------------------------------------------------
```

And each of the lines represents each stage including learning, accumulating knowledge, making mistakes, and more. So Stage 5 is actually the accumulation of all prior stages impacting that particular stage. It's quite dynamic I think.

F.M. – Interesting.

J.T. – I need to do a better job of explaining it! I'm going to re-label my extremely elementary diagram.

```
Elementary school   --------------
High school         -----------------
College             ----------------------------
Graduate school     -----------------------------------------
Postdoc             -------------------------------------------------
```

For me, I'm in the postdoc stage right now. But that doesn't mean that I just forget the fundamentals I learned in elementary school and high school.

F.M. – So, the *x*-axis is the time taken to learn something?

J.T. – Kind of! Maybe this will help if one imagines an alternative framework:

Elementary school   ---------
High school   ----------
College   ----------
Graduate school   ----------
Postdoc   ----------

I disagree with this framework. I think what I'm trying to say (very poorly) is that once you move to a new stage, you don't just suddenly become an expert in the prior stage. You continue to hone those skills. You continue to improve. It's cumulative!

F.M. – Totally makes sense.

J.T. – That was a huge tangent!

F.M. – I know, but good however. Where were we?

J.T. – We were discussing how to teach kids to ask questions!

F.M. – That's right. On that note, Dr. Baden explains it rather nicely:

"Pick a research question that you are passionate about - techniques and skills one can always learn. Surround yourself with people who know more than you. Listen to that weird guy in the corner with crazy ideas - he may be right! Don't be afraid to second-guess textbook knowledge. Science is an iterative process, and only findings that are consistently observed under a range of experimental conditions are the ones that survive the test of time - and sometimes not even those...Oh, and if you don't already, learn at least some basic computer programming - in a few years you won't survive without it anymore in science - don't be left behind."

J.T. – Such fantastic, wise advice! Perhaps that is what we really need to help young people. PASSION!

F.M. – I really do wish we got more responses from our survey. I think there is so much untapped inspiration out there that can be used to empower the next generation of leaders in science.

When asked about the best piece of advice he ever received from a mentor, Dr. Carter writes:

"There are three major factors that contribute to success in science: hard work, intelligent thinking, and luck. But a scientist only needs two out of the three to make it..."

And Dr. Baden? He writes: *"I don't know anything about that - use your brain and go find it out..."*

There is so much wisdom out there. I think our little experiment highlighted just how powerful gathering these can be. We only scratched the surface.

J.T. – I wish there were more responses too, but I absolutely agree. My hope is that these few stories will encourage folks to reflect on their origin stories, and feel empowered to share their thoughts with others!

I never asked you! What is the best piece of advice you ever received from a mentor?

F.M. – Good question! Let's see. It's difficult to choose. All the advice I have received has been useful at one point or another! But mentors have repeatedly told me to enjoy the little wins in life. They matter a lot.

J.T. – The best piece of advice I ever received from a mentor was *"Don't take rejection personally."* I think this includes, of course, paper rejections, grant rejections, and fellowship rejections. But it encompasses failure in its entirety—including failed experiments, particularly repeated failed experiments.

It is really hard for me even now to really pick up the pieces, but I'm actively working on this. I used to take things really hard and really personally but it's not worth it. I remind myself of what this particular mentor told me all the time.

F.M. – Very true. I think such advice applies here too to some extent. We didn't get many responses to our questions, but we didn't get discouraged.

J.T. – Good point! It's really about making the best of what you have. In some ways, I really like rejections from journals particularly if they have comments. They give me a fantastic basis for improvement so I can come back with a punch! Ultimately, I think this piece of advice is so important to me because on the surface it seems quite simple. But the concept has incredible depth. Don't take rejection personally means that you don't dwell or worry. We should seek to always improve, find ways to be resilient, and learn from all our experiences. That's hard to do sometimes!

F.M. – Amen.

J.T. – This reminds me of another thoughtful piece of advice we received from Dr. Audrey Fan who is a postdoctoral fellow at Stanford University in Electrical Engineering. A mentor once told her, *"Bright scientists often fret about whether they can accomplish the tasks and protocol they set out to do. Instead, assume success. You will figure out the kinks and you can make it work. The better question to ask is - who will I help when I am successful?"*

What an incredibly selfless way to think about what it means to be a great scientist. What do you think about this one?

F. M. — This is an important question. It is a question not many people ask of themselves. I think another way to think about this is *how* will I help others when I am successful. I think this is a powerful frame of reference to have. It is difficult to maintain in our current society.

J. T. — We are all guilty of being bogged down in our own self-interests and day-to-day issues, leading us to forget about how we can support others.

This reminds me—I don't know if I ever shared this with you. A while ago, I wrote a mission statement for myself. It encompasses all aspects of my life and is always in flux (I'm constantly editing and tweaking it). But the first part of my mission statement (at least right now!), is the following: To help and support others, lift up those around me, and make meaningful connections with people.

How is indeed an essential question. How do you address it within STEM and education?

F. M. — My personal statement similarly has an outward perspective: to better oneself and those around through service.

With respect to the *how* question and STEM education, I think it partially involves reaching out to the young generation coming behind us. I don't think it requires anything grandiose but simple acts do count as we discussed earlier in the chapter. In fact, they are just as powerful. Consider mentoring a kid in the community. Consider giving a talk at a local elementary or middle school. Consider writing down your thoughts on STEM education and sharing those on social media.

J. T. — I like how you phrased that because I think people sometimes get overwhelmed that they have to do something really big and theatrical. However, the simple acts honestly make the most impact in a straightforward way. Everything is about connections. Dr. Fan mentions this in her one piece of advice she would give to young people interested in science. She wrote to us, *"Surround yourself with people who are intelligent and compassionate, and more importantly, who appreciate what you have to offer to science. The path to discovery is unpredictable at best, and the lasting relationships I've cultivated along the way are as inspiring and significant as the science itself."*

F. M. — It is those relationships that end up mattering most in the end. I would push it to even suggest that those relationships with people you haven't met can be just as important. It is this that I now realize through writing. It has the power of making connections with those you will never meet in person.

J.T. – How do you mean?

F.M. – Let's see. The best way I can explain it is that sharing one's thoughts through writing provides a unique avenue to build connections with others. I just remember the overwhelming feedback I got after publishing *"Tell the Negative Committee to Shut Up"* [5]. I think you can relate to some extent given your recent publications.

J.T. – That absolutely resonates with me. I've been really amazed by how writing can reach so many people. I totally understand what you mean. It's incredibly powerful. I've emailed and communicated with people in other countries about my writing and topics that are of broad interest. It can really bring people together.

Through our website (www.stemadvocacy.org), I've been really trying to fuse these two things: storytelling AND paying it forward to others by mentoring new writers and helping them brainstorm, edit, and go through the process of developing an idea to the final product.

F.M. – Totally. I think this is important especially in the world of science advocacy where stories play a powerful role.

J.T. – Absolutely. I love Dr. Fan's story about how she got interested in science as a child.

"My father grew up in a small fishing village on the coast of Taiwan, his own father a seaman with little formal education. Maybe as a way to create order out of chaos, maybe because he was fascinated by ocean tides while crabbing as a child, my father embraced engineering as his lifeblood. He served in the navy as an engineer to travel the world, studied fluid dynamics for his PhD, and is now a professor of mechanical engineering. I like to think my own itch for adventure, to observe and engineer natural wonders such as our brain and vessels, grew organically from this lifeblood I share with my father."

F.M. – The itch for an adventure! The itch to discover the unknown. Seemed like her father played a key role here.

J.T. – For Dr. Fan, it was clearly contagious! You can tell in her story how her father learned and observed from his father, and in turn, she has also learned and observed from her own father.

F.M. – Passing stories from one generation to the next. I have no doubt she will or is doing the same with those around her.

J.T. – Do you have any good stories from older generations? Ones that have inspired you?

F.M. – Not many actually. It's the stories from my peers and also young people that usually inspire me. Hearing about their experience succeeding or failing. As we have discussed before, it is amazing how high school teachers play a big role in getting young people interested in science.

J.T. – Actually Dr. Minakshi Guha, a senior scientist at BioRad, shares such a story when we asked her how she got interested in science when she was young. She writes, *"It was definitely my high school chemistry teacher who got me interested in science. He was a PhD himself and taught HS chemistry so thoroughly and with passion that I got interested in pursuing science as my major going into college. He was truly a role model for me to follow and I owe it to him for being where I am today."*

F.M. – Just like Dr. Vanderford who was also inspired by his high school teachers. The same goes for me.

J.T. – Which high school teachers influenced you? And more importantly, what did they do specifically?

F.M. – It was my Biology teacher Mrs. Tully. She did something that Dr. Guha shared with us as her piece of advice for young people: *"My advice to young people interested in science would be to keep an open mind and always be curious. Science can help explain so many mysteries of life that it is really endless learning that you can pursue for the rest of your life. Science is so intriguing in so many ways and even after a PhD, I still feel like there are so many things we as scientists don't know."*

Mrs. Tully had a unique way of getting us to think about the questions and the experiments we would need to do to test the hypotheses we generated. I was fortunate enough to be at a school that had the facilities and staff who gave us the space to be curious and ask questions. Of course there were limits, but I vividly remember the many discussions we had in class as we talked about different hypotheses and tried to design experiments to test them.

J.T. – One of my most amazing teachers in high school was Mr. Johnson. He was my Algebra II/Trigonometry and Calculus AB teacher. He had been teaching for over 35 years and even refused to use a textbook. He actually had his own materials that he would give students on the first day of class. They had clearly been written on a typewriter. But, he was the first teacher who really showed me that being smart was cool. He had this little statue of a thinking gargoyle on his desk. And very rarely, when someone answered a question in a clever way or made an insightful comment, or asked a question that stumped him or started a whole new discussion, he would silently walk up to you and place the gargoyle at your desk. I never thought having a gargoyle on my desk would be so satisfying and inspiring but it really was!

F.M. − We really need more great science and math teachers around the world. As we discussed in Chapter 3, The Global Perspective, we must use a multitude of strategies to attract the most talented science students into the teaching profession. In addition to showing that science is cool, teachers can perhaps highlight that teaching is something students can do in their futures.

J.T. − I took for granted that being a great teacher is certainly not easy but it can make such a huge difference. When asked what got him interested in science, Dr. Imeh Williams, now a data scientist, wrote, *"A high school computer science class that focused on computational methods. I remember spending all my time on the assignments and going beyond due to my interest in curiosity. It was one of the few times in high school where I was driven by my intrinsic motivation rather than padding my resume with a good grade."*

F.M. − It is probably the case that Dr. Williams' teachers cultivated and harnessed his motivation for learning. Again, it is clear that it is important to have the right teachers at the right time.

J.T. − Even at an early age!

F.M. − Especially at an early age.

J.T. − Dr. Imeh Williams reiterates this, *"Make sure you find a mentor early on—even as early an elementary school. Everyone needs a champion, someone to show them the ropes. So seek out mentors early and often! I was blessed to have teachers who were able to cultivate my talent and help me navigate the system to ensure that I took advantage of opportunities during the regular school day as well as after school and summer enrichment activities."*

F.M. − It can certainly make a huge difference when you have mentors pushing you forward, both inside and outside of the classroom. Dr. Guha sums it up nicely with the best piece of advice she ever received from a mentor: *"Always keep trying and never give up. Always keep the end goal in mind and strive to achieve it. If you know where you are going, enjoy the journey that comes along with it."*

J.T. − Even if you don't know where you're going, the journey is the best part with all of its meandering, confusion, beauty, success, failure, and most importantly learning.

F.M. – An excellent point. Amidst all this, it is important to remain confident as Dr. Williams reminds us that *"the impostor syndrome is real, especially if you come from an underrepresented group! Don't be an obstacle to your success! There are plenty of challenges along the way so there is no need for you to be one."*

J.T. – Very true. So, we've covered a lot of ground in this chapter on inspiring the next generation. Where do we go from here?

F.M. – Back to the future! But before we do that, I think it is essential to reiterate Dr. Vanderford's inspiring words:

"Dream BIG dreams and pursue those dreams with everything you have. Dream about having an impact on the world by being a behavioral scientist, biomedical researcher, a physician-scientist, a chemical engineer, a bioinformatician, etc. Find individuals (aka, mentors) that will provide encouragement and resources to help you achieve your dreams. Never let a roadblock, including negative people, stand in the way of your dreams. Once you're in a position to do so, do everything in your power to help the next generation achieve their dreams."

Now, let's go to the future!

REFERENCES

[1] NextGen VOICES Results. <http://science.sciencemag.org/content/suppl/2014/04/03/344.6179.34.DC1/> [accessed 01.11.16].

[2] To be or not to be. Stories in Science. <https://storiesinscience.org/2017/02/26/to-be-or-not-to-be/> [accessed 10.03.17].

[3] Mass Mentoring Partnership (MMP). <http://massmentors.org/> [accessed 1402.16].

[4] STEM to STEAM. <http://stemtosteam.org/> [accessed 2802.16].

[5] Muindi F. Tell the negative committee to shut up. Science 2014;345(6194):350.

STEM Education Advocacy Group: A Case Study

F.M. − So, where were we? Ahh yes! The future!

J.T. − The future begins with new ideas!

F.M. − Ideas that will make a difference somehow. Well, at least you hope they will.

J.T. − And often ideas that take us out of our comfort zone.

F.M. − They really should.

J.T. − We have spent much of our spare time working on an idea that you initially had!

F.M. − It seems so long ago! It just started as a discussion with my wife, Moytrayee.

J.T. − Take us back to that discussion! The birth and evolution of an idea is incredibly inspiring. This is the origin story of STEM Education Advocacy Group.

F.M. − The discussion centered around science education, global health, and how to make a difference in developing countries. You see, my wife is in global health. She is used to thinking about issues at a global scale, especially in developing countries. She challenged me to think globally about science and how we could make a difference on a global scale. It was a hard question.

J.T. − An enormous question indeed! So here's another tough question. Once you have an idea, what tips you into doing something? What flips the switch? This, I think, can be really difficult for young people. You can have so many ideas, but how do you take that idea and make it a reality?

F.M. − You are totally right because one could easily get paralyzed since there is a lot that you can and cannot do. At that time, we felt it was important to share our idea with the world actually! This is a little counter intuitive, but we felt that we needed to share our thoughts aloud. For us, this is where it all started. Part of the switch was flipped at that stage. We didn't really have a clear idea of where it would go

Journeys in Science. DOI: http://dx.doi.org/10.1016/B978-0-12-813090-2.00005-1

(the mission statement if you will), but we figured it was important to just start doing something.

J.T. – So, you both started writing!

F.M. – That's right. We shared our thoughts through writing. We wrote a letter to Nature proposing a global fund for STEM education similar to that established for global health. You see, we observed that there really wasn't a central platform to support science initiatives around the world. It was very patchy. So, we thought that perhaps a global STEM fund could be a platform to unite all the efforts around the world in providing more teachers, offering wider access to resources and programs, assessing the success of existing initiatives, and may be even recruiting more women and students from underrepresented communities around the world [1].

J.T. – I imagine it was both exciting and terrifying to put your ideas out into such an open forum. I continue to be astounded by how powerful writing can be in its ability to connect people, even those you have never met, from all over the planet. How did you and Moytrayee feel about putting your ideas out there? What kind of response did you get?

F.M. – Ow yes. Again, at this point, STEM Education Advocacy Group did not exist. We (more me actually) really didn't care if someone else took our idea and brought it to life. What was important was the idea. The overall response to the letter was humbling. We used Twitter and Facebook, of course, to share the idea, and we got great feedback. We were contacted by some organizations that were really excited by the idea, and they were asking us more details about it. Luckily, we got an opportunity to write an additional article explaining the idea in more detail [2]. So, at this point, we started to realize that writing was something that could be used to promote and share ideas around science education. Things had been building up to the point where something (a group or organization) needed to be created to channel the momentum.

J.T. – What other articles did you write? And then, how did you start setting up the group? Seems like a monstrous task.

F.M. – That's right. It looks monstrous when you look at all the little steps together. However, we took lots of little steps. So, there was another article we had written in Nature that proposed the creation of a centralized platform to showcase STEM programs across Africa [3]. This is actually now a reality thanks to Wikipedia [4]. There was also another article in Science, which dealt with the impostor syndrome in graduate school [5].

J.T. – So, at this point, there was not yet an organization but you both had clearly been committed to writing and spreading your ideas! On the verge of something bigger than yourselves...

F.M. – This is important. Again, even without a group or organization, we had started to do the SMALL THINGS consistently. Writing was something that I never thought I would find such pleasure in doing. It was the thrill of sharing powerful and helpful ideas with others.

J.T. – This is a critical lesson for young people. You don't need an organization or a structured group to do something impactful.

F.M. – Absolutely. And that is why we began the book with the famous quote by Lao Tzu: "Every journey begins with a single step." I would add that every journey begins with a single step no matter how small it is.

J.T. – But overcoming the fear is a real challenge for people and much of it is related to the impostor syndrome. We have previously talked about your article on the impostor syndrome. That article was very different from the others. It was much more personal, yes? In some ways, I imagine that may have been easier to write but much more difficult to share.

F.M. – It was very personal indeed. But it fit within the overall mission at that time: this idea of helping trainees in science around the world. What's funny here was that I needed to listen to my own advice in the article where I told my negative committee to shut up [5]. As it slowly became clear that we needed to create a group or organization, the usual doubts started to creep in. *What if it doesn't work? What if no one wants to join? How will I find the time to run the group? What about funding?* The questions were endless.

J.T. – Ah, those doubts. You know what, though? Even now, every time I write something, I still have these thoughts. I don't think they ever truly go away. And I really believe that the reason we have doubt is that we are trying to do something new. If you aren't rocking the boat, there isn't any reason to have doubt or fear.

F.M. – It is important to address some of the questions. You need to talk to a lot of people to get some outside perspectives. Do your homework and see whether you really need to start a group or organization. Perhaps there is already some group or organization doing what you want to do! Why not just join them? Why re-invent the wheel?

J.T. — What was the answer to that question for you? I'm really asking you tough stuff here!

F.M. — This is good! For me, it was actually relatively simple. For me, sharing powerful ideas and analysis through writing was a lot of FUN for me. I get this energy and rush whenever I have an idea for a new article or analysis I want to start doing. Of course, I got challenged especially on the idea that there was no actual ground action. What I mean here is that we were doing more of the thinking (e.g., writing articles) rather than the doing. But of course, I remembered something important: you can't do it all and that you have to start somewhere!

J.T. — Me too! I'm with you! I get so jazzed when I start outlining a new idea. The writing comes very quickly after that. So, what about all the details? Let's start with people. How did you identify people to contribute?

F.M. — Naturally, I went to my friends! Why start an organization if not to work with your friends! I think someone famous said that. Anyway, to answer your question, it was through collaborations. I would propose an idea for an article and I would see whether that individual would be interested. Of course, I would tap them given their expertise and experience in whatever the central thesis was for the article. If there was genuine interest, I would then invite him or her to work on the article and gently hint at the idea of joining the group!

Isn't that how you joined? Gosh, I can't quite remember the moment. Of course we worked together on that Nature Biotechnology article on graduate school mental health [6].

J.T. — Haha! I do remember that you and Moytrayee asked if I would be willing to join! At the time, I was finishing my MD-PhD, but I remember the summer after I graduated, I started writing my own personal piece on my journey through medical school and graduate school [7]. Yes, that mental health piece, I think, really struck a chord with a broad audience.

F.M. — But you are right. There are so many details I could share. Like the numerous rejections to journals (before we started to publish on our website), finding committed people to join, building and maintaining the website (and the social media presence through Twitter, Facebook, and LinkedIn), and of course finding funding. These are good challenges that are still present.

J.T. — These are continued challenges. I think expansion is tough.

F.M. — And determining how much to expand!

J.T. — While maintaining quality and commitment! Gosh, these are tough challenges that I think all organizations face.

F.M. — So for those considering starting an organization in STEM, there is a lot to think through. Again, I would like to suggest that for most, you actually don't need to start your own organization or group. The likelihood to find something already in existence to participate in is very high.

J.T. — I want to emphasize that people often worry about not having enough time. I think this is a legitimate concern. You and I both have our own jobs, and all of the things we write and do for the group and outreach activities, we do on a volunteer basis. But, if you go back to your core values and mission, this in and of itself is enough to motivate you! And, ultimately staying organized is the key.

F.M. — We are all busy. The secret I have found—again—is to do things in small increments. Over time, you are going to be amazed with how much you have done! Of course, being organized is super important.

J.T. — We should share with people a framework that we have for writing that can be applied to any group or organization. It doesn't mean it is what everyone should do. It's just what works for us. The first essential element is to start with a central core thesis. It could be anything that you care about but whatever it is, use this as your crux.

F.M. — The same goes with starting an organization or group. What is your mission statement? Here are some important things to think about: (1) What does your mission statement assume? (2) Does your mission statement inspire others to act on behalf of the organization or initiative? (3) Does it allow you to make important decisions? [8].

J.T. — You should always return to this mission statement if you are lost or confused. It should reflect your core values and serve as motivation. And, importantly, the mission statement should be dynamic. It ought to change with time, and can certainly evolve.

F.M. — Your last point is very important. The mission statement should be dynamic. As such, you should pretty much examine it on a regular basis.

J.T. — So back to writing. Organization is essential. Set reasonable deadlines for yourself but don't beat yourself up if you get busy and can't make it. That's okay! The key thing is to do things in small, scheduled, and regular increments. We try to commit to writing once per week for about an hour. Our mental health article, e.g., was written very quickly in this way [6].

F.M. — The same goes with starting an organization or group. Organization is essential. Tools like Trello, e.g., can be super useful to visualize all the different action items for your organization, and if you are writing, it can also help you stay on track with reaching the different milestones.

J.T. — Another component of organization is creating an outline. Having a skeleton framework sets you up to be more efficient. This is a really simple step that I have found to be SO important.

F.M. — The outline is our lifeblood! Spend time writing a good outline for an article you are writing, your organization structure, and pretty much everything else!

J.T. — We always go back to the outline! Do your due diligence. Just as in science where you start with a literature review before jumping into experiments, you have to do your homework!

F.M. — Of course, doing all these things take TIME! It is worth mentioning that everyone is a volunteer in the STEM Education Advocacy Group.

J.T. — It would be so nice to be able to pay people for their efforts! To get a webmaster and hire someone to run social media too. There is so much to do!

F.M. — All those things require some money! Finding funding for organizations is very difficult. One could try to ask for donations, write grants, or charge for a service.

J.T. — You've done some fundraisers too by running half marathons!

F.M. — Those were fun. In fact, the precursor to this group was Run4Education! This group I started back in graduate school did not survive because its design was fundamentally flawed. It depended on doing runs. I mean a lot of runs. That wasn't sustainable.

J.T. — Indeed! It required quite a bit of running on your part! We have definitely tried and been pretty creative about it now though. But, failure is to be expected. And one certainly should not be discouraged by it. Do you remember the grant we applied for?

F.M. — How can I forget! It was a great experience though.

J.T. — It really was. We were able to write our ideas down in a thoughtful and organized way. And even though we were shot down immediately, we have a nice go-to template now for future grant proposals. I really appreciated that we both laughed it off when we were swiftly rejected!

F.M. — It was a good failure indeed! The Open Philanthropy Project was and still is a unique organization that offers grants across a number of focus areas including scientific research [9].

J.T. — In general, it never hurts to try. I think that's a lesson that we have learned over and over again.

F.M. — The mission statement is very useful when you face such setbacks. And one should expect a lot of setbacks!

J.T. — Indeed! And, we do what we do because we truly believe in the mission statement.

F.M. — The mission and vision statements are even more critical when people are volunteers.

J.T. — What is the difference between a mission statement and a vision statement?

F.M. — For me, the mission statement spells out the what, who, and how of your organization. It is the things that you are doing now as an organization toward your vision. And what is the vision statement? Well, the vision speaks to the "why." Perhaps it is a future state you want the world to exhibit. It is something that may be unachievable (at the moment) but still a worthwhile goal. For us, our vision is *equitable access and exposure to quality science education for every person on the planet.*

J.T. — This distinction is crucial. The vision should spark inspiration and excitement. It should really pack a punch and make you say, "Wow!" I really like our vision.

F.M. — I feel like ours does it. The statement really does energize me to keep going. It speaks to the need to make education and exposure to science universal for every person on earth.

J.T. — For me, when I read the statement, I get completely jazzed and excited. And I can see how everything that we are doing ultimately links back to that vision.

F.M. — We of course hope it inspires other people to help us do some of the initiatives from the group!

J.T. — Definitely! We also can't forget to have fun!

F.M. — This should be close to the top of the list. Do what is fun and also helpful to others at the same time! I think our most recent initiative from our think tank in telling stories in science is something that is fun and helpful to others [10].

J.T. — Stories in Science is such a fun example of ideas that come up in conversation that suddenly become reality!

F.M. – It really is the fun part of our group. Ideas are born through discussions and take a shape of their own over time. The Stories in Science Initiative has really been in the making for quite sometime! I guess the idea was always there but it has now come to fruition.

J.T. – But it has its challenges too. We spent a long while figuring out how to reach people and how to get others to contribute. Honestly, we are still trying to figure that out! But, that's part of the challenge and fun of it all.

F.M. – As we discussed in the introduction, the primary goal of the initiative is to create a place where people can share stories of their journey in science. The belief is that such stories can act as a catalyst in galvanizing the next generation of students in science. However, it truly has been challenging to get stories.

J.T. – It really has been, but that's how we get creative!

F.M. – It has taught us to be persistent. I truly see the initiative taking off eventually.

J.T. – I do hope that it will grow and become commonplace for people to share their stories with one another. I think too often we suffer in silence and we also celebrate in silence. Does that make sense?

F.M. – There is too much silence out there with respect to our experiences in science. I envision the initiative becoming the largest archive of stories in science in the world! Imagine the influence it would have!

J.T. – Imagine if any young person in the world could find and read these stories.

Imagine if one person could connect with another person anywhere on the planet to say, "Wow, I felt exactly the same way" or "You're not alone." How powerful would that be? I think that could become our reality.

F.M. – It is a reality that I believe will come true one day. Science teachers from one corner of the globe could share their stories with teachers and students from another part of the globe. You can imagine all sorts of inspiring stories in science being told that would otherwise never be shared.

J.T. — I get so excited thinking about the global collaboration that could happen. I can't even fathom what amazing things we could produce if we worked together with other people from entirely different backgrounds who have had such a wide variety of experiences. Wow.

F.M. — And it is this aspect of collaboration that our group has embraced just recently.

J.T. — How so?

F.M. — This idea of advising and advocating the many STEM groups out there. Through our projects in mapping STEM programs in Africa and hopefully South America (in the not too far distant future), and our research on organizations here in the United States, it has become evident that there is a lot of value we can add in working with STEM organizations around the world. The idea is to guide and empower science education initiatives and organizations around the world through analysis, advising, and advocacy.

J.T. — And there it is. That is our mission statement. The three As. Each one of them so essential in a unique way.

F.M. — Over time, it became evident that this is where we can make a real impact. It has been so humbling to learn about initiatives all over the world coming up with creative programming to inspire young people and train them in STEM. So, the first A (Analysis) really speaks to the importance of acquiring data.

J.T. — DATA! Can't get enough of it. I think as scientists, we recognize how powerful data can be. And not just haphazardly getting data but acquiring it in a systematic and deliberate way.

F.M. — For organizations, this is very important. You have to be data driven. Seek to analyze everything that you do in a systematic way. Do not assume anything. Have some data to back up your assumptions. Our article on the importance of graduate student mental health was driven by our search for data [6]. As it turned out, we discovered a lack of data prompting us to conclude that there has to be a push for more data acquisition pertaining to graduate student mental health.

J.T. — Precisely. Data gives you incredible leverage. I felt that our search for data really made the mental health article much more powerful. It made our message stronger.

F.M. — And it can certainly make any messaging from an organization or group stronger. So through the first A, we do our own research on important topics in the sciences around the world and of course publish articles in our website and in high profile journals such as Nature and Science.

J.T. – I *always* emphasize the importance of data and citing references when I'm writing and editing. I probably sound like a broken record, but it's really that important.

F.M. – And there are tons of data out there. The trick is in finding the signal from the noise.

J.T. – And what you mention about analyzing science initiatives not only requires data but an organized structure to the data as well. What about the second A, advising?

F.M. – Ah, yes. The advising refers to using our capacity for research (from the first A) to directly advise STEM organizations on their ongoing initiatives. It made sense to do this since it has the multiplier effect. If you are able to influence and help one organization, the idea is that the changes they implement (in response to some of our suggestions) trickle down directly to the students they work with.

J.T. – Ah, yes. It's the incredible cascade effect. It's contagious. I think advising is something that we are actively growing.

F.M. – We literally just started it!

J.T. – I know!

F.M. – The multiplier effect is present in the two As we have talked about (Analysis and Advising). It is even more evident in the third A (Advocacy).

J.T. – Advocacy can really spread like wildfire. We have been lucky enough to advocate through our writing. The Internet really allows those ideas to spread quickly and reach vast numbers of people.

F.M. – As I think more about it, our group is really affecting change through words. We have realized that written words have a lot of power. What is that saying again? The pen is mightier than the sword? Anyway, I think you get the point.

J.T. – That proliferative effect is incredible. One person sends an article to another person, and things can just grow exponentially from there!

F.M. – Something we forgot to mention is the importance of mentorship in writing. We have realized there is a high activation energy that a lot of people face when it comes to writing articles. As we continue to seek new people to join our group, providing mentorship in writing has been something that has been important for us.

J.T. – The multiplier effect definitely has resulted in more interest in writing and analysis from new writers. For me, being very clear and transparent is important. I always try to tell writers that I will help them brainstorm, edit, re-edit, edit again, and re-format until pigs fly. It can take a week or it can take a year, but I will help them produce a solid article that they can be proud of and that is based on strong evidence. As long as they are willing to be persistent and keep working collaboratively!

F.M. – Of course, when you say writers, you mean analysts right?

J.T. – Yes, analysts because, in fact, they are critically analyzing data. And really I think this applies far more broadly than writing.

F.M. – It applies everywhere.

J.T. – How do you handle this mentorship?

F.M. – In our group, it really is about reaching out to people. In fact, this is my approach to mentoring in all domains.

J.T. – I have also found that setting expectations upfront is crucial.

F.M. – Absolutely. Some guidelines are very important. Setting expectations allows each party to know what he/she expects of the other. As a mentor, you have some expectations. As a mentee, you have expectations of the mentor. I think we are setting a nice foundation of mentorship in our group.

J.T. – I agree—we've been able to get some supportive people to help and guide us. It's hard to even remember how we incorporated our current advisors into our group.

F.M. – It was about having someone to discuss some of our ideas with. You can't think in a vacuum. You need some outside perspectives. You need someone to challenge your assumptions.

J.T. – The variety in our mentors has been the most pleasant surprise for me. Each brings a slightly different perspective, a different experience, and a different value.

F.M. – That's exactly right. We owe a lot of thanks to Dr. Nathan Vanderford for his help in getting us on the right path for this book. It is things like this that you want to look to your advisors for assistance. One shouldn't just try to have a lot of advisors for the sake of having advisors. You want the *right* advisors on your team.

J.T. – Team is the key word here. Your advisors should believe in your mission. They should be able to give critical feedback while constantly molding you toward the team's greater vision.

F.M. – I think our group's advisory board is something we continue to build and adjust here and there.

J.T. – Agreed. Our needs may change in the future, and our advisors will ultimately reflect that growth and development.

F.M. − I personally believe that groups must be founded with a mentorship and advising framework. You want people to join because the people in your organization lift each other up and are able to learn from one another through feedback.

J.T. − It's all about the culture of the group. A culture of feedback and openness.

F.M. − A culture of mentorship.

J.T. − I think we have been very deliberate about facilitating a culture of open-mindedness.

F.M. − A culture of support.

J.T. − A culture of collaboration.

F.M. − Yes! Collaboration!

J.T. − A culture of curiosity.

F.M. − In an effort to always improve so that we are better than we were yesterday.

J.T. − So that we can make a positive difference, no matter how small or large, in the world. Here's to a better future!

REFERENCES

[1] Muindi F, Guha M. Developing world: global fund needed for STEM education. Nature 2014;506(7489):434.

[2] Guha, M, Muindi, F. The case for a global fund for science education. <http://www.scidev.net/global/education/opinion/the-case-for-a-global-fund-for-science-education.html> [accessed 11.12.16].

[3] Muindi F, Guha M. Training: African database for education schemes. Nature 2013;504 (7479):218.

[4] List of organizations engaged in STEM education across Africa. Wikipedia. <https://en.wikipedia.org/wiki/List_of_organizations_engaged_in_STEM_education_across_Africa> [accessed 14.03.17].

[5] Muindi F. Tell the negative committee to shut up. Science 2014;345(6194):350.

[6] Tsai JW, Muindi F. Towards sustaining a culture of mental health and wellness for trainees in the biosciences. Nat Biotechnol 2016;34(3):353−5.

[7] Tsai JW. The M.D.-Ph.D. double agent. Science 2015;350(6266):1434.

[8] Muindi F. Mission Statement Matters. STEM Education Advocacy Group 2017. <https://www.stemadvocacy.org/single-post/2017/01/01/Mission-Statement-Matters> [accessed 14.02.17].

[9] The Open Philanthropy Project. <http://www.openphilanthropy.org/> [accessed 09.01.17].

[10] Stories in Science. <http://www.storiesinscience.org> [accessed 08.01.17].

CHAPTER 6

The Future and Beyond

F.M. – So where are we headed? I think a lot about the future of science for those coming after us. What kind of world of science will they inherit?

J.T. – Such a great question. I like to frame this in a specific way. I envision a world of science that is at its core collaborative and kind. I think collaboration amongst individuals, across institutions, across different ages, and across all types of barriers imaginable is what I really hope for science. And when I say kind, I mean that I hope people approach science in an empathetic way toward one another, and that their science is motivated by an innate hope for change, progress, discovery, curiosity, and bringing people together as opposed to pulling people apart.

F.M. – Wow. What a powerful vision. The kind approach as you have described is something that future generations of scientists will have to deeply believe in. It is also my hope that such core beliefs could inspire scientists in the future. But, and yes, there is always a but.

Would the environment support such motivations? Better yet, what kind of an environment is necessary to promote such motivations?

J.T. – Ah yes, there is always a but. I think the environment has to be one where you are safe to make mistakes. And really where any scientist, no matter at what level of training or age, has to constantly feel (1) valued and (2) inspired. I think that it is really easier said than done. What do you think? What qualities are key to promoting those motivations?

F.M. – I agree that a safe space to make mistakes is important for sure. For me, I think a lot about what it means to be successful in science. I think this directly ties into the environment that is necessary to give way to the motivations you mentioned earlier. I naively believe that an environment where the meaning of success is diversified is key.

Journeys in Science. DOI: http://dx.doi.org/10.1016/B978-0-12-813090-2.00006-3

A narrow minded view of the meaning of success within science is limiting.

J.T. – So the key question then is how do you define success, and ultimately I am with you (and I don't think it is necessarily a naive view but one that is a bit more complex). Success for me encompasses so many components including self-worth, impact on others, satisfaction, continued growth and development, leadership, teaching others, mentoring others, teamwork, and community building.

F.M. – Exactly. However, in our current system, I think success is viewed in a myopic fashion in many places. And I think you know what I am talking about...publications. They are easy to quantify and a lot of our awards (heck our scientific enterprise) are built around them. And yes again, let's be clear that publications are important. It is one important way of showing progress. It helps differentiate one researcher from the other. But should it stop there? How can we integrate other measures of success so as to expand the current view? Can we diversify the meaning of success?

J.T. – Great question and I agree publications are indeed important metrics of scientific productivity. But you're absolutely spot on that there are other ways to measure scientific impact. Let's brainstorm what some of these metrics or assays would be!

F.M. – I think mentorship is a good one. However, it is tough to quantify though.

J.T. – Mentorship is high on my list—one proxy for mentorship is seeing what mentees are now pursuing. Are they in leadership positions? What kind of impact are they having in their current roles?

F.M. – I think another measure is to directly assess what students thought about the mentorship they received. Better yet, it is imperative to have some control groups in order to understand the true impact of mentorship on students. I am sure there are surveys that do this already. Anyway, what's next on the list?

J.T. – Community involvement I think is overlooked immensely.

F.M. – Totally. This is something that is important but overlooked. Would be cool to see more labs with an outreach section. I guess there needs to be a way to directly reward outreach activities. Remember though that it takes time away from research. One needs to balance.

J.T. – We seem to get stuck in our academic towers occasionally and forget that we live and work in complex ecosystems and are only one piece of a large community.

F.M. – A very large community.

J.T. – Not just an academic institution or a city but a global community.

F.M. – Here is a part of a lab philosophy that has excited me about the future:

"Our lab is as much a team as it is a family. My job is to coach this team well and make sure all players gel well on and off the court in the service of pushing the scientific ball down the field efficiently and effectively."
Steve Ramirez, PhD

J.T. – That really resonates with me. Wow.

F.M. – It goes back to mentorship. It's all about those relationships you form along the way. That matters quite a bit. This is something someone told me a while ago. People like Steve have a powerful vision of the future and how science can be done.

J.T. – What would be your own lab philosophy? Heck, it doesn't even have to be a lab. What would your team philosophy be? And arguably an even more important question: How do you keep yourself to that philosophy? How do you hold yourself to it? How do you commit to it?

F.M. – My philosophy? One word: Service. I hope to continually develop this in myself and in others. How do I keep myself to that philosophy? By looking for opportunities to serve. Being a mentor for example. Helping others with the words I write. Volunteering once in a while. The trick is to find decent paying jobs where you get to serve and have fun doing it.

J.T. – Service is something that I think is very much overlooked. So much of our society is driven by self-centeredness, demonstrating power, showing dominance over others that ultimately we neglect the importance of caring about other people. We easily forget that there are so many things bigger than our individual selves. I am guilty of this and have to remind myself that I am part of a community far bigger than myself.

I think the word I really mean to use is interconnectedness. I like to think of it as an ecosystem. Do you remember learning about ecosystems in ecology and biology classes growing up?

F.M. – I certainly do.

J.T. – It was always incredible to me to imagine how if one part of the ecosystem was disrupted in even a minor way, this could result in catastrophic and devastating effects for the rest of the community.

F.M. – Sadly, many of us don't think like that. The system is not designed to accommodate "bigger than self" type of thinking. I am just curious how such a philosophy could be integrated into the current scientific system especially for some developing countries which are still building their own scientific systems.

J.T. — This is a complex question to say the least. I'm going to try my hand at it though! For developing countries thinking of their own current scientific systems, I think the root of everything lies in recognizing why science is beneficial for their particular community. Recognizing that STEM education is a means to asking questions, critically assessing infrastructure, and ultimately improving not only the way we live but the efficiency and cost of how we live. I think a strong scientific system has to embrace this concept of science as a core value, if you will. Does that make sense at all?

F.M. — It sure does. Science is a way of thinking much more than it is a body of knowledge. Remember Carl Sagan? I have probably said this too many times now. Anyway this core concept, which I think is what you are talking about, can be useful for developing countries. Of course, you correctly state that this is a complex problem. The core value doesn't fit everyone. Every situation is unique and the priorities are different depending on the circumstances for a given community. However, in my humble opinion, I do agree that science is more a means of asking *good* questions. This should be emphasized early and frequently.

Just an opinion.

J.T. — Definitely every situation is unique but every community has to decide for themselves what the value of STEM is—that is the core foundation that can drive everything. And it's a good value system to return to if things go haywire or are not quite going smoothly.

F.M. — True. The decision part is important.

J.T. — And what I imply by having a "value system," for lack of a better phrase, is that each community has to recognize that science is more than individuals winning Nobel prizes or grants. It is about an underlying current pushing us to think about how science benefits the community. But this current should be infused in all things. It should run through how teams work in science. It should flow (I know the wave analogy is so cheesy) through how people collaborate with one another, how people teach others, and how mentorship occurs.

F.M. — I think a lot about this for sure. What kind of training will students in science be exposed to in the future? What will be fundamentally different to the way training in science is done now for K-PhD and beyond?

J.T. — I'm curious what you think about this. I'm amazed constantly by the pace of technology. We are learning new things every day. But, I actually think that students will be encouraged to think of socially responsible ways of applying science.

F.M. — I am reminded of Stanford's Design School.

J.T. — Do tell!

F.M. — I like their whole concept. Their website describes it pretty well: "In a time when there is hunger for innovation everywhere, we think our primary responsibility is to help prepare a generation of students to rise with the challenges of our times [1]."

They are targeting the process. I think such an approach can force a student to not only think of socially responsible ways of applying science, but of using the process to solve challenging problems in society.

J.T. — Ah I love that. I'm going to ask you a tough question.

F.M. — Let's hear it.

J.T. — Sometimes it seems that technology is driven toward profit instead of social change. Why is that?

F.M. — People need to make money. Money is an easy metric for people to show how successful they are. It sure does go back to the "me" centered mentality. However, not all technology is driven toward profit. Why not do both.

J.T. — True true, they are not mutually exclusive. I do worry (perhaps more than I should) that sometimes people are turned off by social change because it seems, on the face of it all, too challenging. Like something that can't be overcome.

F.M. — They should welcome the challenge.

J.T. — But, in some ways that's why the scientific/engineering approaches are so great. They are all about solving seemingly insurmountable problems!

F.M. — Exactly. The training doesn't give you the answers. Just the tools to figure out the right problem to solve. I feel like there is a theme developing here.

J.T. — What's that?

F.M. — Forget finding solutions. I want problems. Better yet, I want the right problems.

J.T. — Great perspective! What are the "right" problems? What do you mean by that?

F.M. — Tough one. That depends on your topic of research and interests. I think the "right" problem is actually wrong. I think an "important" problem is better. These could be problems whose answers shift the way we think. This is hard though.

J.T. – Ah interesting. And ultimately this is the core of how scientists, engineers, and mathematicians think. When we ask questions, we get more questions. And we learn that it is OKAY to not have answers! And it is more than okay to look for answers.

F.M. – Exactly. It is totally okay not to have answers. I think this is what I envision for the future in STEM education.

J.T. – Maybe that's the crux of everything.

F.M. – Perhaps. So how do we get there?

J.T. – Ah, this is such a tough question. I have so many ideas. And simultaneously I have no idea!

F.M. – It's a tough one.

J.T. – Ahhhhhhh!

F.M. – Let's take it in stages.

J.T. – Okay, let's break it down!

F.M. – Let's say middle to high school. What can be done here?

J.T. – In middle school and even high school, I definitely spent a lot of time learning from books rather than getting my hands dirty. Interestingly, the times I learned the most though were when I was given a super vague task to complete.

F.M. – I think you certainly need to learn from books. The question is how this process should be balanced with question generating opportunities in school.

J.T. – I'll give you some examples. I had a physics class that was mostly learning through fun projects. We had the classic egg drop experiment from the top of the bleachers using minimal supplies. We had another crazy project to build a bridge out of dry pasta that could withstand 100 pounds of weight. It was nuts! It was frustrating! It was fun! It was really crazy.

F.M. – Such experiences are useful. Creating opportunities to generate new knowledge could be useful. This takes a lot of time though. We have to acknowledge that teachers have limited amounts of time. There are standardized exams students have to pass. You only have so many hours in a day.

J.T. – Good limitation to point out. So it's a balance then. Passing the exams and learning the fundamentals in combination with starting to hone creative thinking at the middle school stage. Are there any other ideas you have for approaches to education in middle school?

F.M. – That is not easy.

J.T. – Hahaha.

F.M. – But providing research opportunities could be one way. I know this is being done already but perhaps more of it should be done.

J.T. – Do you imagine these research opportunities to be created by teachers in middle school? Or in collaboration with colleges, universities, and biotech companies?

F.M. – Both actually. One initiative I recently learned about is the Boston Leadership Institute [2]. They offer one and three-week STEM research programs for high school students and K-8 students. The program states that they offer programs that are led by "top teachers from leading high schools with experience teaching high-achieving students." Understandably, the prices are $599 for one week programs and $2200 for the three-week research courses. Of course, this is an example of one program. There are others.

J.T. – Wow, this is so awesome! I'm curious if the cost of the programs prevents students from particular backgrounds from attending. Do they have scholarships or financial assistance available? I do like the concept though of finding great teachers and bringing students to these essentially optimized learning environments. It's a neat concept for sure.

I wonder what other STEM programs there are like this one?

F.M. – In fact, there is a cool website that is connecting all the STEM programs across the United States. It is known as the STEMconnector that has over 6000 STEM-related organizations [3]. It's quite impressive.

J.T. – Incredible. This has the potential to be insanely useful and critical for connecting students to STEM resources. The possibilities for collaboration amongst organizations and sharing data between communities are immense.

F.M. – I think central hubs such as the Connectory and the STEMconnector are super important especially for students. Having a simple way to survey the STEM landscape and find out the right opportunities is very helpful. It is part of the reason why we also released a list of STEM organizations across Africa [4]. The list is incomplete but the idea is to create a central place where students can see all the efforts in the continent. In fact, it is just as useful for the organizations to know what they are each up to. I am a big fan of collaborations.

J.T. − Creating that list of STEM organizations across Africa was an incredible amount of time invested by you! What do you think an ideal collaboration looks like? I think collaborations can be incredibly fruitful if they are done properly, but they can also end up being a waste of time for everyone or merely have a one-sided benefit for a single party.

F.M. − I think it would entail a careful discussion between involved parties. The goals of the collaboration would have to be discussed in great detail up front. Both parties need to know what each other's involvement entails. I think even just having the organizations talk to each other every so often (say a yearly conference or something) would go a long way. Now that I think about it, a conference of STEM organizations across Africa is actually a good idea. Just a random thought.

J.T. − To get back to STEM education in middle and high schools, I'm curious if collaboration is a concept that can be taught through science. Part of the wonder of STEM to me is that the scientific concepts are fascinating, but they are also a conduit by which to learn key skills including collaboration, resilience, communication (verbal, written, graphical representations, PowerPoint for better or for worse), teaching others, resource utilization, analyzing data, and critical thinking.

F.M. − Collaboration is critical in science. I think having students work together in coming up with questions and/or working on experiments would help. I think this is done at some schools and some don't do any of it at all. Of course, access to resources can be a hindrance.

J.T. − I often think that, as students, we don't learn these skills until later in our academic lives—like late in high school or college. I didn't really get a handle on these skills until college, and even now I'm continuing to hone them and improve constantly. Could you imagine how cool it would be if elementary and middle school students started developing these skills early on?

F.M. − Just pure magic!

J.T. − Speaking of these skills, how do you see them developing in creative ways? One of my pet peeves has always been the often stale nature of PowerPoint presentations. What do you think?

F.M. − I say get rid of PowerPoint presentations. Okay, okay. Perhaps not get rid of them but severely limit their use. I think we should embrace going back to the basics (the chalkboard) in getting students to talk about and engage with science.

J.T. — I think PowerPoint is often abused but can certainly be used to communicate very effectively.

F.M. — When used properly.

J.T. — Agreed. I have more recently developed a huge appreciation for the chalk talk or whiteboard talk. Being able to communicate without a huge screen or a computer is actually a skill that I think many people are losing in this age of technology. But, I think it demonstrates mastery of a topic and the capacity to share information with people easily and quickly.

F.M. — I actually think we should start as early as middle and high schools in doing the chalk talks. Why the hell not? In fact, in resource limited settings, the chalkboard is what is often available. Let's use it!

J.T. — Agreed completely. I still remember taking the chalkboard erasers outside during elementary school and hitting them against the ground to get the chalk out! There was always one student in the class assigned to that classroom job every week!

F.M. — Cool!

J.T. — I think another skill that I've reflected quite a bit about is writing. I used to really despise writing. Did you learn the whole five paragraph essay thing? I found that structure so restrictive. It was so prescribed and formulaic.

F.M. — Ow, Yes. I definitely remember the five paragraph spiel. Like you, I used to also despise writing. I think it had something to do with the fact that the writing was forced onto us. You had to write the report. You had to write the essay. I think having additional outlets outside the classroom can also be super useful. I really like the NextGen Voices Essays that Science Magazine publishes [5]. These are typically super short responses to a prompt from the editors. I have seen a publication from a high school student!

J.T. — When did you get interested in writing? How have you developed your skills and how do we teach kids how to write?

F.M. — I got into writing during the last 2 years of graduate school. I realized my passion for writing at that stage. Could I have tapped into it earlier? I think so, but I am happy with the way it worked out. I think the NextGen Voices outlet is one example of a good opportunity for young students to further improve their writing skills beyond the classroom. I would also point trainees to the science communication communities which are very active especially on Twitter! I imagine there are lots of other writing opportunities out there. Blogs can also work. I do think teachers (especially in high school) can encourage students to seek these opportunities early. I think it needs to be in small snippets to slowly get students into writing.

J.T. — What do you think "good writing" is?

F.M. — It depends on what it is you are writing about. Of course having structure and so forth is important. But fundamentally, I think the reader has to be able to follow your logic and connect with the core idea you are trying to communicate.

J.T. — I think what qualifies as "good writing" really depends on who your audience is. But, there are some structural components of writing that are key. Good organization and structure, I think, are essential. I used to write very slowly because I wanted every single word to be perfect. I found it to be incredibly painstaking. I started writing much better when I no longer worried about perfect writing and just started plain, flat-out writing.

F.M. — And that's my advice: Just WRITE!

J.T. — When I started writing more at the end of graduate school, I was blown away by the power of writing. It's so fun to think about how you go from nothing, literally a blank Word document with the cursor blinking, to words on a page. And those words, once they are edited, can be disseminated so widely.

F.M. — And here is the thing I didn't realize until now. Once you do get it published, one's writing lasts forever. Literally forever. It is an interesting way to talk to people throughout time. It is very humbling. I would like to think that people will still be reading this book many many years from now! And I hope the book is of help to them!

But let's bring it back to the future. I thought a lot about this recently during my trip to Kenya where I was reminded of the amazing drive that exists inside young people.

J.T. — Yes! I want to hear all about it!

F.M. — We have talked about this in an earlier chapter. I was fortunate enough to be able to go to support the Harvard Summer School in Kenya that was run by my wife's global health division at the Massachusetts General Hospital. I took advantage of the opportunity (with the help of my own department at Harvard) to also visit local middle and high schools to talk about science. It was my hope to inspire and encourage some of the students into the sciences.

Let me say that I was so impressed with the level of interest that many of the students displayed. I was amazed at the powerful impact that my one little talk had on the students.

I feel a great deal of responsibility to pay it forward.

J.T. – What an incredible opportunity and a privilege to be able to meet young people. I would love to know what you discussed with the middle and high school students? What kind of information did you share with them? What kind of questions did they ask you? What kind of science do they do in school?

F.M. – One thing: Don't be afraid to ask questions.

J.T. – Good because as you can tell, I have quite a number of questions!

F.M. – I told them, "Look around you. . .there is so much we don't understand about the natural world!"

J.T. – Did they embrace this?

F.M. – Beyond what I expected. The brain was hugely popular.

J.T. – This is incredible.

F.M. – I started to wonder how do we nurture this natural curiosity and guide them to ask better and better questions? What kind of schooling would be beneficial to help them think more critically about the natural world they live in?

This is tough in resource limited settings.

J.T. – Well it seems that they are naturally curious already! What does STEM education look like for the kids who you met in Kenya?

F.M. – I can't definitively say but I am willing to bet that it is mostly theoretical. Of course this is likely to vary quite a bit from school to school. I repeatedly questioned them about where they thought information came from in the science books they are reading. I got a wide array of answers. Although schools are limited by the availability of resources, science is still something that is heavily emphasized in classrooms. This was good to see.

J.T. – It seems to me that even simple things like having conversations about science are sufficient to be the initial lightning bolt that spurs young people to become interested in science and, more importantly, give way to more curiosity and questions.

F.M. – EXACTLY. It was this rather simple observation that still keeps me thinking and wondering.

J.T. – So, maybe this is the key—that fancy resources and expensive reagents really are not that important.

F.M. – I really think so. I also think this applies around the world. Young people just need role models to encourage them.

J.T. − Now that you are back from Kenya, have you kept in touch with any of them? Have any of the young people reached out to you? And what kinds of things have they asked or said?

F.M. − Sadly, I have not been able to contact the middle or high schools students I met. This I can understand. It is my goal for the next country I visit to make sure I engage with some students after my visit. However, I have been in touch with several of the university students I met in Kenya.

J.T. − Did many of the kids have laptops or computers?

F.M. − Not in the classrooms. I am sure some students probably have laptops at home. I think we take computer and Internet access for granted here in the United States. It is something that is still a challenge in many developing countries (although it is improving considerably).

J.T. − The truly remarkable point that we last talked about in person was that it only takes one person to be inspired. And they share their inspiration and questions and excitement with their friends. And it spreads virally. The power of knowledge, in this sense, is truly amazing.

F.M. − All it takes is ONE.

J.T. − Precisely. So here's a question for you.

F.M. − Let's hear it!

J.T. − Do you think what you did for a week could be easily replicated amongst more locations? I'm thinking throughout Africa, South America, and Asia. I think the answer is yes!

F.M. − Yes, yes and let's see... yes! Here is what I envision. Let me also say that such efforts are already happening, but are patchy.

The challenge is to get students (middle and high school) some face time with scientists (and also former scientists) across the world. I envision a formal mechanism that provides financial support to scientists, science policy makers, science writers, and many others to travel around the world to share their science, and more importantly, their journey in science. Essentially, they could do a "tell us your story" tour.

What do you think?

J.T. − I think this has so much promise. Ultimately, the underlying issue that we have to discuss is that somehow we have to make the academy more interested in the developing world. How do we make this a priority?

If we could find a core group of scientists, writers, and others to spend a tiny amount of their time doing more outreach in the developing world, we could really instigate a revolutionary process.

And ideally, their experiences would be similar to yours, and would push others to follow suit.

F.M. – There are definitely some members of the academy who are very interested in the developing world. We simply need more scientists to share their stories with young people around the world. As you suggest, getting this core group of scientists to lead the process could be revolutionary.

J.T. – Exactly.

F.M. – Of course, the environment needs to support such activities by scientists who already have many pressures including tenure in academia as an example.

J.T. – What about this same model in the United States? I think it is applicable not solely in the developing world.

F.M. – Very true. I am thinking of a global "tell us your story" program. The same model can apply all over.

J.T. – The next natural question deals with the here and now. What can we do in the meantime? I think the answer is precisely what you have been doing! Going to schools, whether they be in Boston near where we live or even in Kenya, and sharing stories.

F.M. – I think we all can contribute in our own way. Remember, all it takes is one. It doesn't have to take a lot of time. I think a lot of people in science just think they don't have the time. I think you can take advantage of the little time you have to reach that ONE student.

J.T. – And that ONE student may just take those ideas and run with them!

F.M. – Such power!

J.T. – As we always say, it only takes one.

Could you imagine if every scientist just took 1 week out of their life to do this? It would be remarkable. The impact would be immeasurable. Am I being too optimistic, you think?

F.M. – I don't think that's unrealistic. It is 1 week out of their LIFE. I would sincerely hope scientists, and also those trained in the sciences but are no longer at the bench, can take a week out of their lives to pay it forward. In fact, I hope they can do more than that.

J.T. – Agreed entirely. But, let me play devil's advocate. How do we emphasize during training and even at the stage of being a middle or high school student, that outreach is incredibly vital?

F.M. – They need role models. If students see more and more scientists doing outreach, I strongly believe this would slowly inspire young people to do the same.

J.T. – And I think outreach can be super simple. A sixth grader can pay it forward to a fifth grader!

F.M. – Why not? You don't need a Nobel Prize to pay it forward.

J.T. – Yes.

F.M. – Again, like my first question in this chapter, where are we headed? I have a lot of cautious optimism.

J.T. – As do I. There is so much to be hopeful for. I think we have seen that there is an enormous amount of innate potential in young people for asking questions and expressing incredible curiosity.

F.M. – They just need encouragement.

J.T. – Absolutely. I think we both know that science always leads to more questions, and importantly has the potential to truly benefit communities and society as a whole.

F.M. – I am cautiously optimistic because young people interested in science around the world need more support. They need to know that they have the potential to succeed. I don't think the stories they hear should be sugar coated at all. They need to hear about the struggles and the joy. They need to hear it all.

J.T. – My caution is pointed at sharing. I worry that scientists are sometimes secluded, hiding data from one another and competing aggressively with one another for the next high impact paper. We forget that this is a team sport.

F.M. – True. Again, this is a message that needs to be disseminated to young people. Much of science is done through teamwork.

J.T. – And this mindframe of collaboration and sharing, I would argue, actually leads to better science. When we meld such vast diversity of thought, the outcome can only be spectacular.

F.M. – I also want to say that the end goal for young people isn't necessarily to be scientists. It is more about a way of thinking that science instills. That is training that one can take into any field, and more importantly, can now critically look at everyday data with a healthy dose of skepticism.

J.T. – Ah, very good point. I think what we've found in all our reading, thinking, and writing on this topic is that STEM education is not meant to have every person become a scientist. Rather, it facilitates a powerful way of thinking and asking probing questions.

F.M. – This is now the modus operandi of my mind.

J.T. – Explain!

F.M. – Science for me is less about articulating the details of a mechanism but an exercise for the human mind to get better at asking questions. The details are useful but the process of thinking clearly about asking questions is extremely important to me.

J.T. — And if we could take all of these skills: resilience, positive attitude, strong mentorship, humility, logic, curiosity, questioning everything, collaboration, sharing, respect for diversity of thought— and infuse them into the way we live—what a world that would be!

F.M. — Just imagine! So, how do we see this future playing out? There are so many parameters that it is not that useful to spit out predictions. But, I imagine a future where young people all over the world are able to use science as one of the keys for personal growth, social advancement, and an engine to drive global peace.

J.T. — I shared my current mission statement with you in Chapter 4, Inspiring the Next Generation. It is constantly in flux, and I edit it with small nuances almost weekly. Its current iteration is that I see a future in which I can help and support others, lift up those around me, and make meaningful connections with people. I also think about values a lot when I envision the future. And I see a future where young people garner respect, empathy, integrity, diligence, discipline, strong work ethic, compassion, teamwork, and accuracy through their science.

F.M. — I have to say that the journey in writing this book has been very enlightening.

J.T. — We started from a very simple idea of sharing our journey in science thus far.

F.M. — And it's amazing where we have ended up.

J.T. — Pretty awesome.

F.M. — It is my hope that the book inspires the reader. We have taken them on quite a tour.

J.T. — We started from our own personal stories—with you watching airplanes fly by in Tanzania and me playing with the defunct computer chips my father would bring home from work.

F.M. — It's amazing to think that far back. Such stories are important not to forget. For the practicing and nonpracticing scientists reading this book, what should be the main take away point you think?

J.T. — Don't forget your scientific origin story. Reflect often on what amazed you about STEM in the first place. That wonder and awe can fuel you even when facing challenges not only in science but in life. What do you think?

F.M. — I would tell them that young people need to hear your stories in science. These are the stories young people crave. They want to know how you found that one question that got you hooked into science. They want to know the people who pushed you. They want to know what you struggled with. They want to know where you FAILED and what you did about it. Get out there and share this with young people.

J.T. – What is the main take away for our leaders?

F.M. – Continue to find ways to increase investment for the teachers. They need the support. Effective support for K-12 teachers is crucial if we are to make a difference to the student's experience in the classroom.

J.T. – You recently shared this quote with me: "Leaders need to be more like a thermostat and not a thermometer." Leaders can set the tone and encourage those around them to pay it forward. Demonstrating through action is so essential.

F.M. – And finally, what is the main take away for our student readers?

> *J.T.* – Don't be afraid to fail. We are taught to fear failure and to avoid it at all costs. Embrace failure. Fall on your face. But, just remember to stand up and learn from your mistakes. And, importantly, be gracious—there is so much to be thankful for.

F.M. – And don't be afraid to ask questions. The world around us is beautiful. It has so many secrets that people like YOU can uncover. Take a chance and ask that question. You never know where it can take you.

In the words of Carl Sagan, "Somewhere, something incredible is waiting to be known."

J.T. – Here's to the journey! Stay hungry. Stay foolish [6].

REFERENCES

[1] Stanford Design School Website. <https://dschool.stanford.edu/> [accessed 29.03.17].

[2] Boston Leadership Institute. <http://bostonleadershipinstitute.com/research-programs.html/> [accessed 24.05.16].

[3] The STEMconnector. <http://www.stemconnector.org/> [accessed 24.05.16].

[4] List of organizations engaged in STEM education across Africa. Wikipedia. <https://en.wikipedia.org/wiki/List_of_organizations_engaged_in_STEM_education_across_Africa> [accessed 14.03.17].

[5] NextGen VOICES Results. <http://science.sciencemag.org/content/suppl/2014/04/03/344.6179.34.DC1/> [accessed 01.11.16].

[6] Commencement address delivered by Steve Jobs at Stanford University on June 12, 2005. <http://news.stanford.edu/2005/06/14/jobs-061505/> [accessed 17.07.16].

Printed in the United States
By Bookmasters